Occupational Health of Hired Farmworkers in the United States National Agricultural Workers Survey Occupational Health Supplement, 1999

DEPARTMENT OF HEALTH AND HUMAN SERVICES
Centers for Disease Control and Prevention
National Institute for Occupational Safety and Health

Disclaimer

Mention of the name of any company or product does not constitute endorsement by the National Institute for Occupational Safety and Health (NIOSH). In addition, citations to Web sites external to NIOSH do not constitute NIOSH endorsement of the sponsoring organizations or their programs or products. Furthermore, NIOSH is not responsible for the content of these Web sites.

This document is in the public domain and may be freely copied or reprinted.

Ordering Information

To receive documents or other information about occupational safety and health topics, contact NIOSH at:

Telephone: **1–800–CDC–INFO** (1–800–232–4636)
TTY: 1–888–232–6348
Fax: 513–533–8573
E-mail: cdcinfo@cdc.gov

or visit the NIOSH Web site at **www.cdc.gov/niosh**.

For a monthly update on news at NIOSH, subscribe to *NIOSH eNews* by visiting www.cdc.gov/niosh/eNews.

DHHS (NIOSH) Publication No. 2009–119

February, 2009

Foreword

Hired farmworkers form a core component of the agricultural workforce in the United States, numbering an estimated 1.8 million workers. Very little national health data exists on this population because of difficulties in identifying and enumerating them. In 1998, to define the magnitude and scope of hired farmworker occupational health problems, the National Institute for Occupational Safety and Health (NIOSH) collaborated with the Department of Labor to collect occupational safety and health information about a nationally representative sample of hired farmworkers. The collaboration allowed NIOSH to include questions on occupational health in an existing Department of Labor survey, the National Agricultural Workers Survey. The purpose of the original survey continues to be the collection of demographic and employment data on hired crop farmworkers. This document presents a first look at the health data from this collaboration.

This document presents nationally representative data on hired crop farmworker occupational health. Data presented in this document are based on face-to-face interviews with 3,613 hired farmworkers completed between October 1, 1998 and September 30, 1999. Topics covered include musculoskeletal disorders, respiratory symptoms, dermatitis and gastrointestinal problems, pesticide safety training, provision of field sanitation, access to health care, and smoking and alcohol use. Data are displayed for the total population as well as different subsamples of workers based on itinerancy of the workers, years spent working in U.S. farms, the type of crop the farmworker was employed in at the time of the interview, and the number of workers employed on the farm.

This document is an important first step in presenting data on a wide range of health outcomes and potential exposures for hired farmworkers. We hope that it will prove useful for agricultural health and safety professionals, researchers, and farmworker service organizations. The data can be used for program planning, to allocate resources, and to develop interventions that target health problems and barriers to health and develop interventions to prevent injuries and illnesses.

Christine M. Branch, Ph.D.
Acting Director, National Institute for
  Occupational Safety and Health
Centers for Disease Control and Prevention

Executive Summary

National Agricultural Workers Survey (NAWS)

The NAWS is an ongoing national survey conducted by the U.S. Department of Labor (DOL) with the purpose of collecting data on crop farmworkers. Since its launch in 1988, more than 30,000 workers have been surveyed. In response to the Immigration Reform and Control Act of (IRCA) of 1986, the NAWS was commissioned by the DOL to examine shortages of seasonal agricultural services workers while simultaneously observing wages and working conditions. These purposes have since been expanded and now include data collection on household and family composition, employment history, wages, benefits, working conditions, health and safety, housing, income and assets, social services, and immigration status. The survey also collects demographic information specific to farmworkers such as language ability, contacts in nonagricultural jobs, and parental involvement in agriculture. It occasionally includes questions from other agencies with an interest in migrant and seasonal farmworkers. To ensure that different work seasons are accounted for, the NAWS collects information at three different times of the year (see Appendix E). For more information and to order reports, see the DOL NAWS Web site at www.doleta.gov/agworker/naws.cfm.

Occupational Health Supplement

The Occupational Health Supplement was added to the NAWS from October 1998 through September 2002. The NAWS Occupational Health Supplement is a collaborative effort between the DOL and the Centers for Disease Control and Prevention's (CDC) National Institute for Occupational Safety and Health (NIOSH). This collaboration enabled NIOSH to collect health information on a large, nationally representative sample of farmworkers. The NAWS was chosen as the survey in which to include the Occupational Health Supplement because of the innovative methods it uses to reach this population, including the following:

- The use of culturally literate interviewers appropriate to the population
- Enumerating and contacting farmworkers at the worksite
- Considering seasonal and geographic employment fluctuations in the design of the sampling plan

The Supplement's main purpose is to obtain national prevalence data on variables related to the occupational health of farmworkers. Topics covered in the occupational health supplement and reported in this document include:

- Pesticide safety training
- Pesticide handling and personal protective equipment
- Field sanitation
- Musculoskeletal pain or discomfort
- Skin conditions
- Respiratory symptoms
- Gastrointestinal problems
- Doctor diagnosed health conditions
- Cigarette and alcohol use
- Quality of and access to health care

Executive Summary

An occupational injury supplement was also added to the NAWS in October 1998, and those results will be presented in a separate report. The overall number and rate of injuries by age group can be found in chapter 3 of the NIOSH Chartbook (http://www.cdc.gov/niosh/docs/chartbook/). McCurdy and Carroll [2000] also present data on agricultural injury.

After the first year of data collection, October 1998–September 1999, some questions were removed from the occupational health supplement to lighten the burden on farmworkers participating in the survey. Therefore, to present the data in a consistent manner, only the data from the first year are presented in this publication.

Highlights of the Data

Study Design
Data presented are from a cross-sectional survey of 3,613 employed farmworkers in 54 county clusters throughout the continental United States between October 1, 1998 and September 30, 1999.

Demographic/ Work Characteristics (Tables 3-4)
- The average age of farmworkers was 31 years, but for those with less than 5 years of farm work it was 25 years.
- Farmworkers were mostly male (78%).
- Nearly 50% of farmworkers were settled and did not migrate for work.
- Approximately 40% of the farmworkers worked on medium sized farms with 11 to 50 total farmworkers.
- Fruit and nut crops employed approximately 40% of the workers.

Pesticide Safety Training (Tables 5-8)
- A third of farmworkers had not received any pesticide safety training in the last 5 years.
- A fourth of those who reported receiving training said that their training consisted of informal instructions in the field.
- 5% of farmworkers trained said that the pesticide safety training they received was not in their primary language.
- For workers who reported receiving training, 11% reported that the training did not cover one or more of the following points: how soon they could enter a field after it was treated with pesticides; illnesses or injuries due to pesticides; and where to go or whom to contact for emergency medical care.

Executive Summary

Pesticide Loader, Mixer, Applicators and Personal Protective Equipment (Tables 9-12)

- 11% of farmworkers report that they have loaded, mixed, or applied pesticides. This varies widely depending on years of farm work, migrant status, farm size, and crop type.
- The most common type of personal protective equipment that was worn the last time they loaded, mixed, or applied pesticides was a suit (69%), followed by goggles (66%). Thirteen percent of loader/mixer/applicators do not wear any type of gloves, and 18% wear cloth gloves.

Drinking Water, Toilets, and Hand Washing Supply Availability (Tables 13-16)

- 78% of farmworkers reported they had drinking water and cups available every day.
- 86% reported that they had toilet and toilet paper available every day.
- 77% reported that they had hand washing water, soap, and towels available every day.
- Workers with more years of farm work, workers on farms with fewer workers, and workers in field crops were less likely to have water and cups; toilet and toilet paper; and water, soap, and towels.

Health Symptoms in Last 12 Months (Tables 17-20)

- Musculoskeletal pain or discomfort was the most commonly reported health problem of farm workers (15%). Almost 20% of farmworkers with 10 or more years of farm work reported pain or discomfort in one or more body parts.
- 6% of farmworkers reported back pain or discomfort.
- Settled farmworkers reported respiratory symptoms more often than migrant farmworkers.
- Farmworkers on farms with more than 10 workers reported musculoskeletal and respiratory symptoms less often than farmworkers on farms with fewer workers.
- 14% of farmworkers reported respiratory symptoms (runny stuffy nose or watery itchy eyes).
- 7% reported dermatitis, most commonly affecting the hands and arms.

Smoking and Alcohol Use (Tables 21-24)

- 25% of farmworkers were current smokers (within the last 12 months).
- 28% of farmworkers with 10 or more years of farm work were current smokers.
- 50% drank alcohol during the month before the interview.
- 57% percent of farmworkers with 10 or more years of farm work drank alcohol during the month before the interview.

Executive Summary

Access to and Quality of Health Care (Tables 25-28)

- Only 36% of farmworkers had used any health care services in the United States in the last 2 years.
- 4% of farmworkers who had used health care services in the last 2 years reported that they sought care for a job-related matter.
- 51% of farmworkers said that health care was difficult to obtain in the United States.
- 41% of farmworkers had never seen a dentist.

The Audience

There are many individuals and institutions that may have an interest in the findings of this document. Researchers interested in the health of farmworkers and clinicians who care for migrant and seasonal farmworkers will find this document to be useful. It will also be of interest to local and national organizations that serve farmworkers and the migrant clinic network.

Program staff and administrators will be able to use this information in a variety of ways. They can use it to plan interventions to target the health problems (Tables 17–20) and health behaviors (Tables 21–24). Through the information provided in this document, they can identify those who do not have health care, and who feel that health care is difficult to obtain (Tables 25–28). In addition, data will be useful for policymakers who are interested in the safety and health needs of this special population.

Table of Contents

Disclaimer ii
Ordering Information ii
Foreword iii
Executive Summary v
    National Agricultural Workers Survey (NAWS) v
    Occupational Health Supplement v
    Highlights of the data vi
    The Audience vii
Abbreviations xiii
Acknowledgements xiv

Chapter 1: Orientation 1
    Significance of the problem 1
    The NAWS Occupational Health Supplement 3
Chapter 2: Strenghts and limitations 5
    Strengths 5
    Limitations 6
Chapter 3: Methodology 9
    Time frame for data collection 9
    Population 9
    Sampling 9
    Data collection 10
    Development of the NAWS Occupational Health Supplement 10
    Data analysis 13
    Participation rates 14
Chapter 4: Results 17
    Part One: Demographics 17
    Part Two: Data from the Occupational Health Supplement 24

List of Appendices

Appendix A: Survey Instrument A1
Appendix B: Questionnaire location for items in tables B1
Appendix C: Crops reported by farmworkers in Spanish
                       and English and crop categories C1
Appendix D: Organizations represented in questionnare planning meeting for
                       NAWS Occuapational Health Supplement D1
Appendix E: Detailed sample selection process E1
Appendix F: Definitions of terms F1

Table of Contents

List of tables

Table 1.	Priority occupational health outcomes for hired farmworkers	10
Table 2.	Reason for indeterminate eligibility of growers	15
Table 3.	Means, proportions, and standard errors (SE) for demographic and work characteristics of farmworkers	19
Table 4.	Demographic variables by years of work on U.S. farms	23

Participation in pesticide safety training programs by

Table 5.	Years of work on U.S. farms	27
Table 6.	Migrant status	28
Table 7.	Number of farmworkers employed on farm	29
Table 8.	Crop category	30

Personal Protective equipment worn by pesticide loaders, mixers, or applicators by

Table 9.	Years of work on U.S. farms	32
Table 10.	Migrant status	32
Table 11.	Number of farmworkers employed on farm	33
Table 12.	Crop category	33

Availability of drinking water, toilets, and hand washing facilities by

Table 13.	Years of work on U.S. farms	35
Table 14.	Migrant status	35
Table 15.	Number of farmworkers employed on farm	36
Table 16.	Crop category	36

Health conditions and symptoms by

Table 17.	Years of work on U.S. farms	39
Table 18.	Migrant status	40
Table 19.	Number of farmworkers employed on farm	41
Table 20.	Crop category	42

Smoking and Alcohol Use by

Table 21.	Years of work on U.S. farms	44
Table 22.	Migrant status	44
Table 23.	Number of farmworkers employed on farm	45
Table 24.	Crop category	45

Table of Contents

Access to and quality of health care by
Table 25.	Years of work on U.S. farms	48
Table 26.	Migrant status	49
Table 27.	Number of farmworkers employed on farm	50
Table 28.	Crop category	51

Physician diagnosed health conditions
Table 29	Estimated prevalence of physician diagnosed health conditions	52

Table of Contents

List of Figures

Figure 1. Grower participation in NAWS 16

Figure 2. National orgin of farmworkers 17

Figure 3. Hired farmworkers by migrant status 18

Figure 4. Ethnicity of farmworkers 18

Figure 5. Immigration status of farmworkers 18

Figure 6. Mean age of farmworkers by years of U.S. farm work 22

Figure 7. Percentage of workers without legal authorization to work in the U.S. by
years of U.S. farm work 22

Figure 8. Pesticide safety training 24

Figure 9. Percentage of workers teporting no pesticide safety training any time
during the last five years by crop category 25

Figure 10. Percentage of workers whose pesticide safety training consisted of
informal instructions in the field by Years of U.S. farm work 25

Figure 11. Language of training 25

Figure 12. Percent of workers who received pesticide safety training in their primary
language by number of farmworkers employed on farm 26

Figure 13. Percentage of workers reporting training that covered three WPS required
topics, by years of U.S. farm work 26

Figure 14. Farmworkers trained in the last 12 months who received a certification
card for pesticide safety training, by years of U.S. farm work 26

Figure 15. Percentage of farmworkers who loaded, mixed or applied pesticides in
the U.S. in the last 12 months, by number of farmworkers employed on farm 31

Figure 16. Use of personal protective equipment, by years of U.S. farm work 31

Figure 17. Availability of drinking water, toliets, and hand washing facilities by
Number of farmworkers employed on farm 34

Figure 18. Musculoskeletal pain/discomfort and dermatitis symptoms by number of
farmworkers employed on farm 37

Figure 19. Respiratory symptoms by number of farmworkers employed on farm 38

Figure 20. Percentage of farmworkers who are current smokers and percentage who
consumed alcohol in the last month, by crop category 43

Figure 21. Health care access by migrant status 46

Figure 22. Use of health care servicesin the United States in the last 2 years, by
number of farm workers employed on farm 46

Figure 23. Employment paid health care not related to farm work, by years of farm
work 47

Abbreviations

BLS	Bureau of Labor Statistics
CDC	Centers for Disease Control and Prevention
CoA	Census of Agriculture
DHHS	Department of Health and Human Services
DOL	Department of Labor
EID	Education and Information Division
EPA	Environmental Protection Agency
ETA	Employment and Training Administration
FLS	Farm Labor Survey
FSS	Field Sanitation Standard
IRCA	Immigration Reform and Control Act
JTPA	Job Training Partnership Act
NAWS	National Agricultural Workers Survey
NCFH	National Center for Farmworker Health, Inc.
NIOSH	National Institute for Occupational Safety and Health
NSC	National Safety Council
OSHA	Occupational Safety and Health Administration
PPE	personal protective equipment
PPS	probabilities proportional to size
QALS	Quarterly Agricultural Labor Survey
REI	restricted-entry intervals
se	standard error
SIC	Standard Industrial Classification
SOII	Survey of Occupational Injuries and Illnesses
UI	unemployment insurance
USDA	United States Department of Agriculture
WPS	Worker Protection Standard

Acknowledgements

The authors of this publication:

Andrea L. Steege
Sherry Baron
Xiao Chen
National Institute for Occupational Safety and Health

Would like to acknowledge our research partners:

Daniel Carroll,
U.S. Department of Labor

Susan Gabbard and Vanessa Barrat,
Aguirre International

Richard Mines,
California Institute for Rural Studies

We would also like to thank the following:
- Farmworkers who participated in NAWS;
- Farm operators who agreed to participation of workers in this survey;
- Jorge Nakamoto, Alberto Sandoval and Aguirre International interviewers who collected these data and participated in the piloting and refinement of the Occupational Health Questionnaire;
- Steve Reder, NAWS Statistical Consultant;
- John R. Myers, NIOSH Division of Safety Research;
- Martin R. Petersen, NIOSH Division of Surveillance, Hazard Evaluations, and Field Studies;
- Aaron Geraci, Greg Hartle and Pam Schumacher, desktop publishing;
- Aaron Geraci, Greg Hartle, Pam Schumacher, Sally Toles, Donna Pfirman, production, design, and logistics;
- Susan Afanuh and Dyani Saxby, editing;
- Geoff Calvert, Paul Middendorff, John Sestito, Sharon Cooper, Don Villarejo, and Stephen McCurdy—Reviewers;
- National Center for Farmworker Health and Focus groups who commented on a draft document at the 2003 Midwest Farmworker Stream Forum;

Special thanks to John Sestito, Toni Alterman, Marie Haring Sweeney, and Bill Eschenbacher for their support of this project.

Chapter 1: Orientation

The Department of Labor (DOL) estimates that approximately 1.8 million workers perform hired agricultural crop work in the United States [DOL 2000].

For the remaining sections of this report, the term "farmworkers" will be used to describe workers performing crop agriculture [all crops included in the 1987 Standard Industrial Classification (SIC) code 01][1]. As defined by the U.S. Department of Agriculture (USDA), crop agriculture includes "field work" in the vast majority of nursery products, cash grains, and field crops, as well as in all fruits and vegetables. The NAWS also includes those who work in the production of silage and other animal fodder [Mehta et al. 2000].

Significance of the problem

Agricultural Hazards

It has been well documented that agriculture is a hazardous industry [Merchant et al. 1989; Brackbill et al. 1994]. Agricultural workers in general are often exposed to hazards that cause injury or ill health, including the following:

- Chemicals that may have long or short-term health affects [Beaumont et al. 1995]
- Plants that may cause allergic reactions [Ballard et al. 1995]
- Heavy or awkward tasks that take their toll on the body's musculoskeletal system [Schenker 1996; Villarejo and Baron 1999]
- Livestock and machinery that may cause debilitating injuries including noise induced hearing loss [McBride 2003]
- Injuries or illness resulting from exposure to the elements [OSHA 1992]

Lack of National Data

Agriculture has consistently been ranked among the most dangerous industries in the United States. For example, in 2005, while the fatality rate for all industries in the United States was four per 100,000, the fatality rate for agricultural workers was more than seven times as high (32.5 per 100,000 workers) [Bureau of Labor Statistics (BLS) 2005]. In spite of the many hazards associated with agricultural work [Beaumont et al. 1995], few national data exist on the occupational health of farmworkers [Villarejo 2003].

Two major reasons for this are reliance on home addresses to locate participants in a population that is often migratory and living in unconventional housing [Sherman 1997] and the use of information provided by employers rather than workers, which can lead to inaccuracies due to underreporting [Leigh et al. 2001]. For example, the BLS national Survey of Occupational Injuries and Illnesses (SOII) uses employer-provided data [BLS 2002], which can be limiting in studying farmworkers as many jobs are short term and farmworkers might not feel comfortable reporting injuries or illnesses to their employer.

---

[1] 1987 SIC codes have now been replaced by 1997 NAICS (North American Industry Classification System) codes. The NAICS code that currently includes farmworkers in the NAWS is "111."

Chapter 1: Orientation

Occupational Health Implications for Farmworkers

Hired farmworkers, because of types of work they perform, and the lifestyle related to this work have characteristics that distinguish them from most other farmers or farm family members and may put them at greater risk either directly or indirectly [Meister 1991]. These include workplace and housing hazards, organization of work, and cultural and community factors.

Workplace and housing hazards

Farmworkers often perform specialized repetitive tasks that require prolonged stooping, overhead work, and heavy lifting with incentives (piece rate) for working quickly and without breaks [Villarejo and Baron 1999]. Even those who do not work on a piece rate basis may perceive that there is a threat of being fired if they work too slowly [Austin et al. 2001].

Farmworkers may have close contact with chemicals either directly through loading, mixing, or applying them or indirectly through drift or residues [Arcury and Quandt 2001]. Exposures may be aggravated by a lack of facilities for washing hands and clothing [Austin et al. 2001].

Housing proximity to fields may allow pesticide drift to enter the housing facility and surroundings. Farmworkers may also bring home pesticides through contamination of their work clothes. In effect, the workers are subjected to dual exposure, both on and off the job [Fenske 1997; Moses et al. 1993; Bradman et al. 1997]. Living in migrant housing may put farmworkers at risk for infectious diseases because of poor water quality, higher rates of tuberculosis, and other infectious diseases [Ciesielski et al. 1991, 1992].

Organization of work

Farmworkers often work for labor market intermediaries or labor contractors. These contractors often "recruit, hire, train, supervise, and dismiss" farmworkers as a crew and supply the crew to farm owners or operators as they are needed. By doing this, it is possible for farm owners to shift responsibility for regulatory and migration issues to the contractor. The contractor is the farmworkers' main contact and authority figure [Villarejo and Baron 1999].

Cultural and community factors

Cultural factors also exist that may put farmworkers at risk. Since we know from previous years of National Agricultural Workers' Survey (NAWS) data collection that many of the workers are Latino, and a large number are also immigrants, they may be stigmatized in the communities they work in, either for their ethnicity or simply for being strangers [Hovey and Magaña 2002, 2003; Alderete et al. 1999]. This may cause increased stress and ill health in the workers [Shuval 1993].

In addition, because many farmworkers are new to the United States, they may be unaware of laws that are in place to protect their health. Likewise, they may be unaware of hazards that threaten their health and safety [O'Connor 2003].

Because of their migratory lifestyle, many farmworkers also experience a loss of social support, which is exacerbated by the fact that many also leave spouses and children behind. A growing trend is minor farmworkers who are often unaccompanied by their parents [Mines et al. 1999].

Chapter 1: Orientation

In addition to loss of social support, moving for work may also mean that a worker is unfamiliar with services available in the place they are working, which may result in an inability or hesitance to seek preventive or even necessary health care [Weathers et al 2004].

Anecdotal accounts have suggested that farmworkers and children of farmworkers drink from, wash clothes, or bathe in irrigation ditches or runoff ponds that may be contaminated with pesticides, chemicals, and organic wastes [NCFH 1985–2002; Meister 1991].

Low English literacy may have implications for health if workers cannot read and understand warning signs, instructions, educational pamphlets, other safety materials, or even express concerns over an employer's use of pesticides and safety [Austin et al. 2001].

The issue of immigration status can have potential health ramifications. Lack of legal immigration status may affect farmworkers' access to health care services, as well as decisions about issues such as the following: questioning the safety and health practices of their employers, seeking medical care, joining labor unions, and making housing decisions [Villarejo and Baron 1999].

The NAWS Occupational Health Supplement

As described in the previous section, many factors can put the health of farmworkers at risk. Because these are often specific to the farmworker population and differ from many other worker populations in the United States, difficulties in studying farmworkers are intensified. Yet, the dangerous nature of agricultural work demands that we overcome such difficulties to effectively investigate the occupational health of farmworkers. The NAWS Occupational Health Supplement was developed as a step toward accomplishing this and as a way to surmount some of the major limitations of other national surveys that include farmworkers. The need for the supplement, as well as its purpose and development will be discussed in this section.

Need for the Supplement

Special circumstances must be considered in the study of the occupational health of farmworkers. For example, some farmworkers do not work year-round or are employed day-to-day. Another concern is that data collected from employers may not accurately measure the occupational health of farmworkers because of underreporting [Leigh et al. 2001]. The NAWS Occupational Health Supplement was developed as a solution to such problems in surveying farmworkers. Its methods overcome several obstacles by obtaining information from the worker rather than the employer and by not depending on respondents having permanent U.S. addresses.

What is the NAWS?

The NAWS is an ongoing national survey of farmworkers conducted by the U.S. Department of Labor with the purpose of collecting demographic and economic data on crop farmworkers. Since its launch in 1988, over 30,000 workers have been surveyed. In response to the Immigration Reform and Control Act of (IRCA) of 1986, the NAWS was commissioned by the DOL for the purposes of examining shortages of seasonal agricultural services workers, while simultaneously observing wages and working conditions. These purposes have since been expanded and now include data collection on household and family composition, employment history, wages, benefits, working conditions, safety and health, housing, income and assets, social services, and immigration status. The survey also collects demographic information specific to farmworkers such as language ability, contacts in nonagricultural jobs, and parental involvement in agriculture. It occasionally includes questions from

Chapter 1: Orientation

other agencies with an interest in migrant and seasonal farmworkers. To ensure that different work seasons are accounted for, the NAWS collects information at three different times of the year (see Appendix E). For more information and to order reports, see the DOL's NAWS Web site at: www.doleta.gov/agworker/naws.cfm.

Purpose of the NAWS Occupational Health Supplement

The NAWS Occupational Health Supplement was added to the core NAWS to better understand the relationships that might exist between these demographic and economic and occupational characteristics and adverse health outcomes. However, because of the cross-sectional nature of the NAWS and the Supplement, etiologic studies may be needed to confirm exposure-disease associations. Nevertheless, there may be more immediate opportunities for intervention regarding health outcomes through training, use of engineering controls, and personal protective equipment. The NAWS Occupational Health Supplement serves as a useful tool for identifying problems that merit further investigation and possible intervention. This project was undertaken to provide the first nationally representative data on the occupational health of hired farmworkers.

Chapter 2: Strengths and Limitations

Researchers have noted that farmworkers are a difficult population to study for various reasons including the following:

- Their tendency to move for employment
- Their nontraditional housing
- Undocumented working status in the United States
- Cultural barriers

Consequently, they may fear confiding in strangers. Through the careful methodology of the NAWS, a number of difficulties in studying the farmworker population were overcome, but some still lingered and new challenges arose. The strengths of the NAWS, followed by its weaknesses, both over time and specific to the year October 1998 through September 1999, will be discussed in this section.

Strengths

National Statistical Sample

The NAWS is the only survey of farmworker health with a national population-based sample of hired crop farmworkers. No other national survey includes a sufficient number of hired farmworkers while simultaneously employing a sampling strategy that accounts for geographic and seasonal fluctuations, as well as farmworker lifestyle. Other more geographically limited surveys have enumerated all farmworkers residing in a given area such as a housing development, county, or city and then selected a random sample of farmworkers. This is an appropriate method for smaller surveys, but for a national survey it would not be feasible.

Sample Based on the Workplace

The NAWS sample is based on the workplace; most farmworkers employed during a given year have the opportunity of being selected for the survey.

Strategic Methodology

The NAWS was created specifically to survey farmworkers. Methods were developed for sampling and surveying this population with consideration for their special circumstances. Specific techniques are employed to overcome cultural barriers. Top priority is given to ensure that interviewers are both culturally and linguistically competent. The primary languages of the farmworkers interviewed were Spanish (87%) and English (11%). Previous national data provided by the NAWS has indicated similar percentages of farmworkers whose primary languages were English and Spanish [Mehta et al. 2000]. As a result, the NAWS interviewers are either monolingual Spanish speakers or bilingual Spanish-English speakers and have had previous knowledge or relationships with farmworkers. Hence, they not only understand the Spanish language, but also the language of farmworkers. All new interviewers are mentored by experienced NAWS interviewers and take part in an extensive 2-day training workshop.

Chapter 2: Strengths and Limitations

Information Obtained Directly from Farmworkers

The NAWS surveys farmworkers directly, in face-to-face interviews. Other surveys collect data on farmworker occupational health through secondary sources, such as the employer. This is the case with the BLS Annual SOII, which requires employers to keep a log of work-related incidents. However, this method has a weakness in that the employer may be unaware of the extent of his or her workers' work-related health problems, especially if they work through farm labor contractors or on a day-to-day basis. Underreporting of illness and injuries by the BLS may occur for other reasons as well [Rosenman et al. 2006; Leigh et al. 2001]. A strength of the NAWS is that the information is reported directly by the farmworkers about conditions they face and their own illnesses and injuries. A two-part study of California farmworkers that included both a survey and a separate medical work-up found that agricultural workers are reliable sources of information on their own health problems (Villarejo, 2000).

Arrangement of Questionnaire

To set the farmworker at ease, the questionnaire is arranged with the least intrusive questions at the beginning of the interview, such as those pertaining to work history and family composition. More sensitive inquiries including those related to health are located in the latter part of the questionnaire. The final question deals with the highly sensitive issue of immigration status. (See Appendix A).

Workers Choose Where Survey is Administered

A major strength of the survey is that the workers are able to choose where the survey is administered. They have the option of completing the survey in the privacy of their own home or at another location, which reinforces confidentiality and may also alleviate fear of reprisals.

Comparable to Other Studies

Efforts were made to select standardized questions, when possible, for use in the NAWS so that the results could be compared with those of other studies.

Limitations

One Year of Data

For the purpose of this document, only the first year of the survey (October 1998 through September 1999) is presented. The length of the survey for the first year differs from that of subsequent years and would thus make comparisons of the data highly complex. Still, many of the questions pertaining to health from the first year's survey were preserved in sequence in successive years' surveys to facilitate the ability to examine trends. The first year had the most extensive occupational health section. Data from the latter years will be included in future documents.

Sample Not Large Enough to Examine Sub-Groups

Since there is a very small percentage of non-Latino hired farmworkers, it is not possible to separately analyze racial/ethnic groups of farmworkers in the NAWS. More regionalized studies that focus on locations with higher numbers of racial/ethnic groups other than Mexican-born, Latino farmworkers are needed to analyze the occupational health of other groups as separate populations. Furthermore, a very small percentage (19% for the first year) of the workers surveyed is women, limiting the reliability of comparison of male and female farmworkers.

Chapter 2: Strengths and Limitations

Age Factor Makes Comparisons Difficult

Farmworkers participating in the NAWS had a median age of just 29 years compared with all workers in the United States having a median age of 40 years [Di Natale 2002]. This age discrepancy may be due to the strenuous physical demands of farm work. In effect, the discrepancy makes it especially important to standardize ages to compare this population with other workers.

Healthy Working Population

Because this is a workplace survey and all farmworkers are required to have worked at least one of the last 15 days in agriculture to qualify for the survey; it is a working population. The "healthy worker effect" is a phenomenon in which the working population is generally healthier than the total population [Last 1995], which includes workers, those people who are not working because of mental or physical incapacity, and those who are not working for other reasons. Farmworkers who have been injured or become ill and have not worked for at least 1 of the last 15 consecutive days are not eligible to participate in the NAWS. The healthy worker effect is expected to be especially strong in farmworkers because of the physical demands of their jobs [Hernberg 1992]. The healthy worker effect is not restricted to agricultural workers but is a factor in studies of workers in other occupations and industries as well. One other consideration is the "healthy migrant effect," which hypothesizes that the healthiest and strongest members of a population are the ones who choose to migrate [Franzini and Ribble 2001]. This may add to the healthy worker effect of those participating in the NAWS since more than half of farmworkers migrate to obtain work [Mehta 2000].

Interrelation of Variables

The results of the NAWS Health Supplement are presented in this document using four types of stratification variables: (1) farm size, classified by the number of workers employed on the farm; (2) type of farm, classified by the crop category; (3) experience in farm work, classified by the farmworkers' years of U.S. farm work; and (4) migrant status. Although useful in describing these results, many of these variables are interrelated. For example, an obvious relationship exists between less than one year of work on U.S. farms work and being classified as a "newcomer" for migrant status. Data will continue to be analyzed and modeled for further clarification of associations and presented in later publications.

High Percentage of Undocumented Workers

Previous years of NAWS data collection show that a high percentage of farmworkers do not have legal authorization to work in the United States [Mehta et al. 2000]. This might have dissuaded the workers from agreeing to be interviewed for the NAWS. A review of interviewer records for a 14-county sample from 1999 showed that 76% of workers who were asked to participate in the survey agree to take part (see participation rates page 21).

Indigenous Languages

Only two percent of those surveyed reported primary languages other than Spanish (87%) and English (11%). However, concerted efforts were made to find someone in the community such as a family member or friend to translate in such cases where the primary language was not Spanish or English.

No Corroboration of Data

The information is self-reported and is not corroborated by medical examinations, medical records, or testing.

Chapter 2: Strengths and Limitations

Cross-Sectional Survey

Since the NAWS examines a specific population during a specific time, it is an indication of the prevalence of certain conditions at the times the surveys are administered [Last 1995]. Since follow-up investigations with the same farmworkers are beyond the scope of the NAWS, some cause and effect relationships cannot be known.

Non-crop Farmworkers Not Included

Because the NAWS was specifically mandated by Congress to survey crop workers, it does not survey those farmworkers employed on other types of farms, such as livestock farms.

Estimated Size of the Farmworker Population

The NAWS does not independently estimate the size of the farmworker population and instead uses a fixed estimate of 1.8 million workers [DOL 2000]. Although this does not have any impact on the percentages reported in this document, it could affect the reliability of estimates of the total number of farmworkers affected by health outcomes derived from the tables.

Chapter 3: Methodology

Time Frame for Data Collection

The data presented in this publication were collected in Federal fiscal year 1999, which began in October 1998 and ended in September 1999. Data are based on interviews with 3,613 farmworkers. Some questions were discontinued after the first year of data collection (October 1998–September 1999). As a result, only data from the first year are included in this publication. Subsequent data will be included in future publications.

Population

The NAWS is a survey of workers aged 14 or older performing crop agriculture [all crops included in the Standard Industrial Classification (SIC) code 01][1]. The definition of crop work by the USDA includes "field work" in the vast majority of nursery products, cash grains, and field crops, as well as in all fruits and vegetables. Crop agriculture also includes the production of silage and other animal fodder [Mehta et al. 2000]. The NAWS population consists of nearly all farmworkers in crop agriculture, including field packers and supervisors, and even those also holding non-farm jobs. Ranch, greenhouse, and nursery workers are also included, so long as they perform crop work that is included in the definition above. However, the survey excludes livestock workers. It also excludes secretaries, mechanics, H–2A temporary farmworkers (nonimmigrant, alien workers permitted to work on a seasonal or temporary basis to ensure sufficient workers for employers and to protect U.S. jobs and wages) [USDA 1988] and unemployed agricultural workers. Farmworkers who have not worked in agriculture at least one day in the 15 days before being asked to participate in the survey are ineligible for the survey.

Sampling

The NAWS collects data on a national random sample of U.S. crop workers that is designed to be sensitive to regional and seasonal fluctuations in labor usage. Each State in the continental United States is in 1 of 12 regions. The NAWS was designed to account for seasonal fluctuations that are characteristic of the agricultural work force by having three interviewing cycles, which last 10–12 weeks each and start in February, June, and October. The number of interviews allocated to each cycle and region varies and is dependent upon the amount of crop activity during a particular season as estimated using data from the BLS and the Census of Agriculture (CoA) [Mehta et al. 2000]. Respondents are selected using a multistage sampling method. The probability that a farm will be selected increases or decreases based on the size of its seasonal agricultural payroll [Mehta et al. 2000]. For a more detailed explanation of the sampling strategy see Appendix E.

---

[1] 1987 SIC codes have now been replaced by 1997 NAICS (North American Industry Classification Sytem) codes. The NAICS code that currently includes farmworkers in the NAWS is "111."

Chapter 3: Methodology

Data Collection

After the farms are selected, the NAWS interviewers contact the growers, describe the purpose of the NAWS, and ask permission to enter the work site. All growers have the right to refuse to allow the interviewers onto the work site. With the growers' consent, interviewers go to the farm (or ranch or nursery), describe the purpose of the survey to the workers, and choose a random sample of workers to participate. Once they agree to participate, the workers choose a time and place for the interview to be conducted [Mehta et al. 2000].

Development of the NAWS Occupational Health Supplement

In 1995, the National Institute for Occupational Safety and Health convened an expert panel on hired farmworker occupational safety and health. The panel issued an official report in 1998 that recommended new directions for surveillance of farmworker occupational safety and health [Wilk 1998]. The priority areas identified in this report are shown in Table 1. Using these recommendations as a starting point, NIOSH convened a 2-day working meeting in spring 1998 to develop the questions for the October 1998–September 1999 Health Supplement to the NAWS. The meeting was attended by researchers from government agencies, community organizations, and research agencies who are experts in farmworker health (see Appendix D). One of the lead NAWS interviewers also participated to provide insight into issues related to the target audience and the way the interviews are conducted. During this meeting, participants first prioritized the key outcomes to be measured by the survey and then met in small working groups for in-depth discussion on each topic. Following the meeting, participants provided further suggestions for question formats and wording. Whenever possible, standardized questions were chosen. Health-related questions from previous years of the survey were retained to examine trends.

Table 1. Priority occupational health outcomes for hired farmworkers*

Outcome
Musculoskeletal disorders
Pesticide-related conditions
Traumataic injuries
Respiratory conditions
Dermatitis
Infectious disease
Cancer
Eye conditions
Mental health

*NIOSH Workgroup on Priorities for Farmworker Occupational Health Surveillance and Research, May 5, 1995

A draft questionnaire was then developed and reviewed by the core interview staff. It was translated into Spanish and pilot tested in 1998 with working farmworkers in several regions of the country. Following the pilot testing, a meeting was held with NIOSH researchers and NAWS field staff to review the pilot test results and revise the questionnaire. A second set of pilot tests and final revisions followed. Once the questionnaire was finalized, a 2-day training session was held with the core field interviewer staff. Since the NAWS had never included health-related questions in the past, this training was important to address interviewers' inquiries and concerns about these questions. Following the first NAWS cycle of 1999, modest alterations were made based on input from the field interviewers. Although questions on neurological symptoms and violence were included in the final version of the questionnaire, interviewers felt that they were misinterpreted and disrupted the flow of the questionnaire. It is possible that the questions were formatted in a manner that was too sensitive or had another meaning for the farmworkers, so that they did not take the questions seriously, or they did not understand the questions. As a result, it was decided not to include data from the sections on neurological symptoms and violence in this report.

Federal Regulation of the Agricultural Workplace

To understand the working conditions of farmworkers, some of the questions in the NAWS Occupational Health Survey were based on two of the Federal standards that regulate the agricultural workplace. These are the Environmental Protection Agency's (EPA) Worker Protection Standard (WPS) and the Occupational Safety and Health Administration's (OSHA) Field Sanitation Standard (FSS). The following descriptions are summaries of the sections that pertain to data in this report. The complete text of these regulations can be found at the following Web sites:

EPA's WPS: http://www.access.gpo.gov/nara/cfr/waisidx_08/40cfr170_08.html

OSHA's FSS: http://www.access.gpo.gov/nara/cfr/waisidx_08/29cfr1928_08.html

EPA's WPS

The EPA's WPS is a regulation aimed at reducing the risk of pesticide poisonings and injuries among agricultural workers and pesticide handlers. The WPS contains requirements for pesticide safety training, notification of pesticide applications, use of personal protective equipment, restricted entry intervals following pesticide application, decontamination supplies, and emergency medical assistance. Because of its complex nature, only the provisions pertinent to questions in the Occupational Health Survey will be summarized here.

Pesticide safety training: Pesticide safety training is required for all workers and handlers. Agricultural employers must assure that untrained workers receive basic pesticide safety information before they enter a treated area on the establishment. No more than 5 days after their initial employment has begun, all untrained agricultural workers must receive the complete WPS pesticide safety training. The agricultural employer must also ensure that the training is delivered to workers in a manner they can understand. Employers are given the option of training their workers and handlers themselves, or hiring workers who have already been trained. In either case, employers must ensure that their employees understand the basic concepts of pesticide safety. Workers and handlers must be retrained every 5 years.

Emergency assistance: Employers are required to notify workers of the location and phone number of the nearest medical care facility or provider to be contacted in the case of pesticide poisoning or injury emergency. They are also required to ensure that the worker is provided with transportation to that medical care facility if a worker or handler may have been poisoned or injured. Information must also be provided about the pesticide to which the person may have been exposed.

Restricted-entry intervals (REIs): REIs are the time period after application of a pesticide when worker entry into the treated area is restricted. They are specified on all agricultural plant pesticide product labels. Employers are required to inform any worker who may come near a treated area

Chapter 3: Methodology

either orally or by posting warning signs. Workers are excluded from entering a pesticide treated area during the restricted entry interval, with only minor exceptions.

Personal protective equipment (PPE): PPE must be provided and maintained for handlers and early-entry workers. Requirements for PPE are based on the toxicity category of the formulated product.

Decontamination supplies: Decontamination supplies (soap, water, paper towels) must be available when a worker enters an area treated with pesticides and will contact a treated surface. These supplies are for routine washing and as well as emergency decontamination. For pesticides that have REIs longer than 4 hours, supplies must be maintained for 30 days after the REI expires. Decontamination supplies are required for seven days following the REI if one or more low-risk pesticides have been applied. Low-risk pesticides are defined as pesticides with REIs of 4 hours or less.

The WPS took effect on October 20, 1992. Revisions were made to training requirements January 1, 1996 and to decontamination supply requirements in June 1996. Since the data in this report were collected after these dates, these revisions would have already been in effect. For more information, see the following Web sites:

www.epa.gov/pesticides/health/worker.htm

www.epa.gov/oppfead1/safety/workers/trainreq.htm

Occupational Safety and Health Administration's Field Sanitation Standard

The Occupational Safety and Health Act of 1970 was enacted to ensure safe and healthful working conditions for working men and women. In 1987, OSHA issued regulations establishing minimum standards for field sanitation in agricultural settings. The OSHA FSS requires employers, who employ 11 or more field workers on any one day during the previous 12 months, to provide the following:

- Potable drinking water
- Toilets
- Hand washing facilities

Toilet and hand washing facilities: One toilet and one hand washing facility is required for every 20 workers. The facilities must be located within a quarter mile walk, or if this is not feasible, at the closest point of vehicular access. Such facilities are not required for employees who do field work for 3 hours or less each day, including travel to and from work. The definition of hand washing facility includes an adequate supply of potable water, soap, and single use towels. Likewise, a toilet facility includes provision of toilet paper, adequate to worker needs.

Drinking water: Agricultural employers must also provide potable drinking water, suitably cool and in sufficient amounts, dispensed in single-use cups or by fountains, located so as to be readily accessible to all workers.

Chapter 3: Methodology

The standard was expected to reduce heat-related deaths and injuries, urinary tract infections, and exposure to agrichemicals and agrichemical residue.

The standard took effect May 30, 1987 for potable drinking water, and July 30, 1987 for toilets and hand washing facilities.
(www.dol.gov/esa/whd/regs/compliance/whdfs51.pdf;
www.safetyinfocur.com/factsheets/OSHA9225.html)

Data Analysis

Data presented are simple prevalences of exposures, health outcomes, provision of sanitary facilities, use of PPE, pesticide safety training, as well as standard errors for these estimates. In addition to an overall prevalence, we looked at the data from four different perspectives. Two of these reflect the situation of the farmworkers themselves, and include migrant status and years in U.S. farm work. The other two reflect the worksite, and include the number of employees on the farm and crop category. Stratifications were not meant to imply causation, but only to describe the data. More sophisticated analyses of the data are being carried out and may enable us to clarify the relationships between these variables and the health data.

Multiple logistic regression analyses were performed to assess the association between all dependent variables with two categories and the four classification variables, including years working in farm work in the United States, migrant status, number of farmworkers employed on farms, and crop categories. Multiple linear regression models and generalized multinomial logit models were applied to continuous dependent variables and dependent variables with more than two categories, respectively.

Satterthwaite-adjusted Chi-square statistics were used to test whether any prevalence differences exist among different levels of each classification variable. All tests were two-sided with statistical significance defined as $p<0.05$. SUDAAN statistical software was used because of the complex survey design of the NAWS [Babubhai 1997]. Data were weighted to estimate population means and prevalence. Variances were estimated assuming with replacement sampling and employing the Taylor series linearization method. The prevalences were calculated using the weight supplied by DOL.

Standard Error

The proportions reported in this document are based on a sample (see Appendix E) of the farmworker population. The deviation of a sample estimate (in the case of this report, percentages and means) from the value that would have been obtained if the entire population had been studied is called the standard error (se). The se of an estimate is a measure of the variation among the estimates from the possible samples and, therefore, is a measure of the precision with which an estimate from a sample approximates the average result of all possible samples. In other words, se is a measure of the accuracy of a given estimator. In this publication, the estimator is the prevalence (i.e. percentage of workers trained, percentage of workers with musculoskeletal discomfort or pain) or in some cases the mean of a variable (i.e. age, highest grade).

Chapter 3: Methodology

Participation Rates

For any survey, it is important to determine the participation rate of the potential respondents. There are many different reasons why potential respondents may not participate in a survey, including lack of interest, lack of time, fear of reprisals, or some other reason. When participation rate is low and the cause for nonparticipation is related to the outcomes being measured by the survey, then the survey results may be "biased." This means that the findings may either overestimate or underestimate the true conditions of the target population being studied. For example, those farmworkers who have experienced a health problem may be more motivated to fill out a survey about health problems than healthy workers. Another scenario may be that farmworkers who work for employers with poor working conditions may fear reprisals by the employer and may not participate.

The NAWS is particularly complicated with regard to measuring participation rate, because poor participation can be an issue at each phase of the complex multistage manner in which the sampling is conducted. The growers have the right to refuse participation and must first agree to allow the interviewers on their property before any farmworker can be invited to participate. As a result, the participation rate should be examined at the grower stage followed by the farmworker stage.

Participation rates for the NAWS survey was estimated through an in depth analysis of participation rates in a random sample of 14 of the 54 participating county clusters and is described in detail below. In summary in those 14 counties, a list of 259 potential grower participants was developed of which 53% were successfully contacted and invited to participate. Of those growers invited, 71% agreed to participate. Once a grower has agreed to participate, a random sample of farmworkers employed by that grower is invited to participate in the study. In those 14 counties, 76% of invited farmworkers completed the survey. It is difficult to determine the impact of nonparticipation on the results presented in this report. If growers with less favorable working conditions avoid participation or if farmworkers fear reprisal and either underreport concerns or do not participate, the results could underestimate the true prevalence of health effects or adverse working conditions.

Grower Participation Rate

To date, the approach to constructing grower lists has been as inclusive as possible. The backbone of the list is data on growers in the SIC code *Crop Production* who participate in the unemployment insurance (UI) system. Since UI eligibility in agriculture differs from other industries and by State, coverage varies dramatically. California and Washington have near universal coverage while many states exclude small farms. In addition, one or two States have historically refused to supply UI data to the NAWS. The UI list, while a good start, does not provide sufficient information. In addition, the information is a year or so out of date. UI data is supplemented by a variety of techniques including obtaining local lists, reviewing local directories and seeking information from knowledgeable persons (e.g., extension agents). The quality of this additional information varies from excellent to poor.

To evaluate participation at the stage of the grower selection, the grower lists in 14 randomly selected counties from the 1999 NAWS were reviewed (See Figure 1). Overall, in these 14 counties, there were 468 growers on the lists. Growers were randomly ordered and interviewers attempted to contact them and interview farmworkers in that order. When they completed the allotted number of interviews for that county, they stopped contacting growers and moved on to the next sampled county. In these 14 counties, 259 growers were contacted randomly. Interviewers were asked to make two attempts to contact a grower and to use a variety of means (phone, in person) and to try different dates and times.

Chapter 3: Methodology

This first step in the contact process was to determine whether the grower was eligible to participate in the survey. The major reasons for ineligibility were as follows:

(1) Not being an active farm
(2) Not currently employing farmworkers
(3) Being engaged in livestock rather than crop production

In this sample, the most common cause for a grower not to participate was the interviewers' inability to contact the grower to determine eligibility. Table 2 shows the reasons why eligibility could not be determined in 122 (47%) of the 259 growers. Some of these reasons could be interpreted as refusals, such as phone calls not being returned, but others are less clear, such as not being able to locate the address of the grower.

Table 2. Reason for indeterminate eligibility of growers

Reason	Number of growers reason applies to
Incomplete contact, nobody around	28
Incomplete contact, answering machine	22
Incomplete contact, person unavailable	13
Unable to determine, unable to locate address	20
Unable to determine, address out of county — unsure if fields in county	12
Unable to determine, (other)	27
Total	122

Of the remaining 137 growers who were contacted, 87 (64%) were determined to be eligible, suggesting that as many as one-third of the growers on the grower lists may be ineligible. Of these 87 growers, 62 (71%) agreed to participate in the survey, but interviews were only completed at 47 growers (54%).

There are various reasons why interviews may not be completed even when the grower has agreed to participate. Very few growers can accommodate the interviewer on the day they are first approached. Most of those who cooperate ask the interviewer to come back in a few days to better accommodate work schedules or for other reasons such as a foreman or ranch manager needs to obtain approval from the owner. In some cases, growers are contacted later in the visit because they are further down the list, and the interviewer never returns because he has already met his quota of interviews.

Chapter 3: Methodology

Figure 1. Grower participation in NAWS

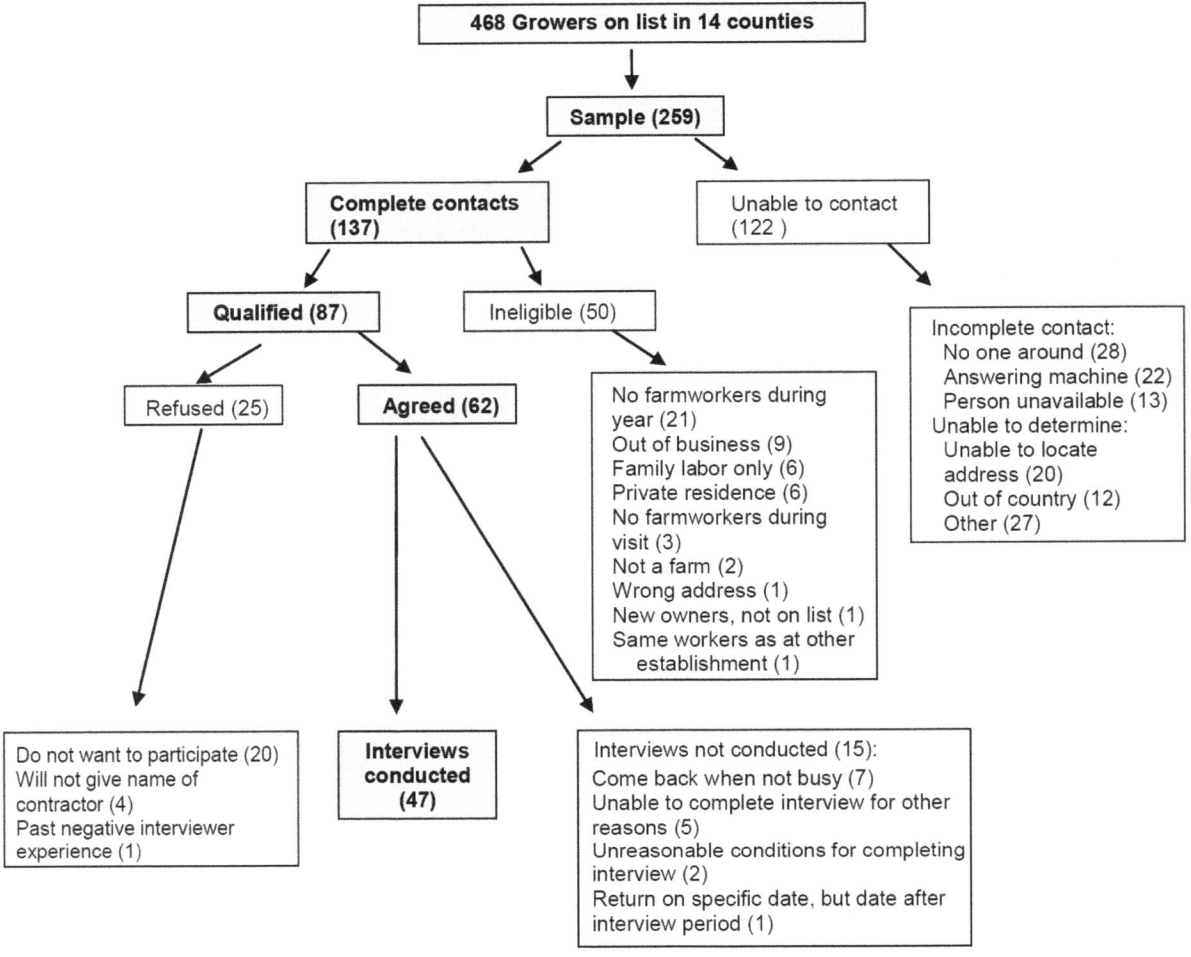

Farmworker Participation Rate

Participation rate of farmworkers was also determined through this same review of interviewer records from October 1998 through September 1999. Although full cooperation was obtained from 47 growers, sufficient interviewer data necessary to determine participation rate of farmworkers was only available for 29 growers. The omission of the necessary data appeared to be random interview error and, thus, it should not result in biased estimates. These 29 eligible growers had 261 farmworkers, and 199 (76%) of them participated in the study. We do not know why the other 24% refused to participate.

Chapter 4: Results

Note. Corresponding data tables are located at the end of each section.

Part One: Demographics

Section One: Summary

This section summarizes demographic variables of the farmworker population surveyed in the Occupational Health Supplement. For the reader's convenience, complete data regarding demographic variables from the NAWS is located in Table 3 in Section 2. For a fuller discussion of demographic and workplace characteristics, see the DOL NAWS research reports: www.doleta.gov/agworker/naws.cfm.

National origin

Of the 1.8 million workers performing hired agricultural crop work in the United States, more than four-fifths (84%) were foreign-born. Nearly all (97%) of these foreign-born workers were born in Mexico. See Figure 2.

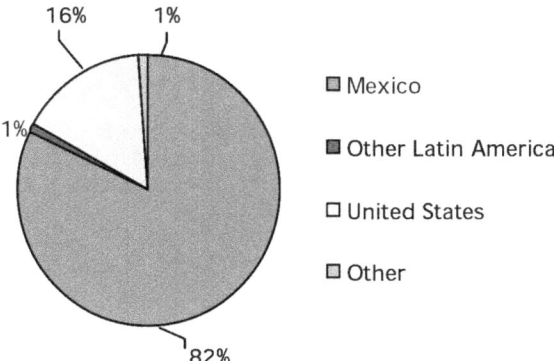

Figure 2. National Orgin of Farmworkers

Migration (Table 3): The three exclusive migrant types used in these analyses are as follows:

Newcomer

A farmworker who was born outside the United States and entered the United States in the year preceding the interview. Excluded from this category were workers who had any farm-work, non-work, or non-farm work period in the United States for 12 months or more preceding the interview.[1]

Follow-the-crop farmworker

A farmworker who has had more than one U.S. farm work job and the jobs have been more than 75 miles apart. This assumes that they would have to establish a temporary domicile at or near the second job site. Follow-the-crop farmworkers can be either U.S.- or foreign-born.

Shuttle farmworker

A farmworker who moves once for agricultural employment during the year then returns to a "home base" to live for the remainder of the year and may work at some other job but *not* in agriculture. (If they did work in agriculture, they would be considered "follow-the-crop"). Shuttle farmworkers can be either U.S.- or foreign-born.

Approximately one-half (51%) of the farmworkers reported that they migrate, meaning they were newcomers, follow-the-crop, or shuttle migrants. The remaining 49% of the farmworkers were settled (see Figure 3). Of the workers who migrate, 34% reported leaving family members behind, including spouses and children.

---

[1] According to the work grid. See page A-8, Survey Instrument.

Chapter 4: Results

Figure 3. Hired Farmworkers by Migrant Status

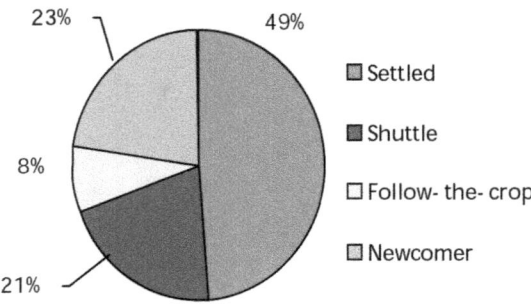

Ethnicity

The majority of farmworkers were Latino (88%) and of Mexican heritage (86%) (see Figure 4). Since there are very few non-Latino hired farmworkers, it will not be possible to analyze racial/ethnic groups of farmworkers in the NAWS separately. Nevertheless, non-Latinos may be excluded to determine whether health effects are more pronounced among Latino farmworkers. Other more regionalized studies may be needed to analyze the occupational health of racial/ethnic groups other than Mexican-born, Latino farmworkers as separate populations.

Figure 4. Ethnicity of Farmworkers

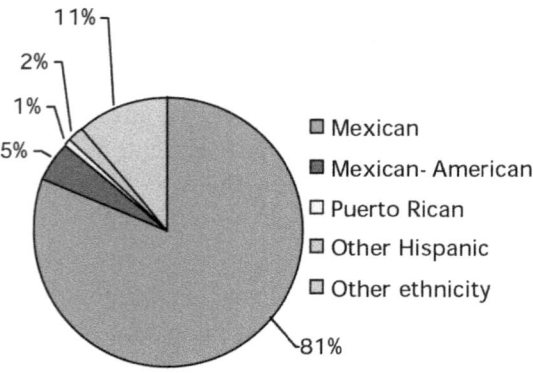

Gender

The farmworkers were predominantly male (78%). Since only 22% were female, we opted not to stratify the data by sex.

Age

The median age of farmworkers was 29. The mean age was 31 and ranged from 14 to 75.

Income

More than half of farmworkers (54%) had yearly earnings (in the United States) below poverty level. This excludes farmworkers who were not present in the United States for the whole previous calendar year.

Immigration status

A large proportion of farmworkers (53%) did not have legal authorization to work in the United States. See Figure 5.

Figure 5. Immigration Status of Farmworkers

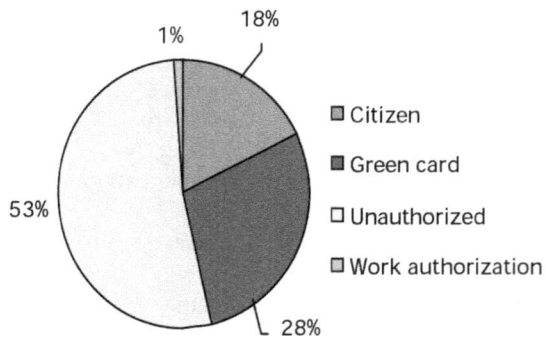

Chapter 4: Results

Section Two: Demographic and Work Characteristics of Farmworkers

Table 3 describes the demographic and work characteristics of hired farmworkers surveyed from October 1, 1998 to September 30, 1999. These data come from the core NAWS questionnaire and are presented so that the demographic data are available for the same year the health data are provided.

Table 3. National Agricultural Workers Survey means, proportions, and standard errors (SE) for demographic and work characteristics of farmworkers, October 1998—September 1999

Variable	Mean (se)[1]	% (se)[1]
Age	31.4 (0.7)	
Male sex		78.0 (2.5)
Foreign born		84.0 (2.8)
Years in the United States (for those foreign born)	8.5 (0.7)	
Place of birth		
Mexico		81.7 (3.1)
Other Latin America		1.4 (0.6)
United States		16.0 (2.8)
Other		0.9 (0.5)
Race		
White		53.0 (2.9)
Black/African American		4.6 (1.6)
American Indian/ Alaska Native/Indigenous		6.8 (2.0)
Asian/Native Hawaiian/Pacific Islander		0.7 (0.4)
Other		34.9 (3.0)
Ethnicity (Hispanic)		
Mexican		80.6 (3.2)
Mexican-American		5.0 (0.9)
Puerto Rican		1.0 (0.3)
Other Hispanic		1.7 (0.4)
Other Ethnicity (non Hispanic)		11.1 (2.7)
Family status		
Nuclear family member lives in household		40.5 (3.1)
Marital status of farmworker		
Married		53.7 (1.9)
Separated/divorced/widowed		5.2 (1.2)
Single		41.1 (2.3)
Children		
Children in household	0.8 (0.1)	
Nonresident children less than 18	0.4 (0.0)	
Total children	1.2 (0.1)	
Family composition		
Farmworker is a parent		48.1 (2.2)
Farmworker lives with parents		11.3 (1.4)
Farmworker married but does not have children		2.0 (0.5)
Other		38.6 (2.2)

[1] (se) – Standard Error. continued

Note. Due to rounding, some column totals may not add up to exactly 100 percent.

Chapter 4: Results

Table 3. National Agricultural Workers Survey means, proportions, and standard errors (SE) for demographic and work characteristics of farmworkers, October 1998—September 1999 (continued)

Variable	Mean (se)[1]	% (se)[1]
Language		
Primary Language		
Spanish		87.1 (2.9)
English		10.6 (2.7)
Other		2.3 (0.6)
Ability to read English		
Not at all		60.5 (3.7)
A little		21.7 (1.9)
Somewhat		4.3 (0.7)
Well		13.6 (2.6)
Ability to speak English (for those whose primary language is not English)		
Not at all		49.1 (3.7)
A little		27.9 (1.8)
Somewhat		7.0 (1.2)
Well		16.0 (2.9)
Education		
Highest grade completed	6.8 (0.2)	
Participation in adult education		21.6 (2.5)
Income		
Family income below Federal poverty level		54.1 (3.4)
Percentage of farmworkers by Family income categories (U.S. earnings)		
<$500		16.7 (3.2)
$500–$999		1.9 (0.8)
$1,000–$2,499		5.0 (1.2)
$2,500–$4,999		7.0 (1.3)
$5,000–$7,499		9.3 (1.6)
$7,500–$9,999		13.2 (1.7)
$10,000–$12,499		11.3 (0.9)
$12,500–$14,999		8.5 (0.8)
$15,000–$17,499		6.6 (0.6)
$17,500–$19,999		4.6 (0.6)
$20,000–$24,999		7.2 (1.4)
$25,000–$29,999		3.8 (0.7)
$30,000–$34,999		3.2 (1.2)
$35,000–$39,999		0.7 (0.2)
>$40,000		1.2 (0.4)
Immigration status		
Citizen		18.3 (2.8)
Green card		27.5 (2.5)
Unauthorized		53.3 (3.6)
Work authorization		0.9 (0.3)
Legal application		
Legalization applicant		14.8 (1.6)
Family program		10.3 (1.6)
Other authorization		5.4 (2.0)
Unauthorized		53.3 (3.6)
Citizen by birth		16.3 (2.8)

continued

[1] (se) – Standard Error.

Note. Due to rounding, some column totals may not add up to exactly 100 percent.

Chapter 4: Results

Table 3. National Agricultural Workers Survey means, proportions, and standard errors (SE) for demographic and work characteristics of farmworkers, October 1998—September 1999 (continued)

Variable	Mean (se)[1]	% (se)[1]
Work characteristics		
Years in farm work	8.7 (0.6)	
Hourly wage	$6.47 ($0.12)	
Number of weeks spent abroad	10.9 (1.4)	
Number of weeks doing farm work in the United States	27.1 (1.0)	
Number of weeks doing non-farm work in the United States	3.4 (0.7)	
Number of weeks not working in the United States	7.8 (0.6)	
Hours worked last week in farm work	41.0 (0.9)	
Employer		
Grower		72.1 (6.2)
Farm labor contractor		28.0 (6.2)
Method of payment		
Hourly		79.8 (3.4)
By piece		15.2 (2.6)
Salary		2.8 (1.3)
Combination of hourly and by piece		2.3 (1.0)
Equipment expenses covered by		
Grower/contractor		69.0 (4.9)
Farmworker		11.4 (3.0)
Farmworker pays some		7.6 (1.8)
Equipment not needed		11.1 (3.2)
Other		1.0 (0.4)
Housing		
Farmworker rents from non-employer		57.6 (4.8)
Employer provides free housing for farmworker		13.9 (3.1)
Farmworker owns the house		17.2 (2.9)
Farmworker rents from employer		2.1 (0.7)
Employer provides free housing for farmworker and his/her family		4.5 (1.4)
Farmworker rents from government or other institution		3.2 (1.0)
Farmworker receives free housing from government or other institution		0.5 (0.3)
Method of transportation to work		
Carpool		38.1 (3.0)
Drive car		35.9 (2.9)
Labor bus		18.2 (4.0)
Public transportation		0.4 (0.3)
Walk		6.8 (1.9)
Other		0.6 (0.2)

[1] (se) – Standard Error.

Note. Due to rounding, some column totals may not add up to exactly 100 percent.

Chapter 4: Results

Section Three: Demographic Variables Stratified by Years of Work on U.S. Farms

Table 4 presents demographic variables by years of U.S. farm work to demonstrate where workers fall within the stratifications presented throughout the rest of the tables. The following is a summary of the demographic variables as they relate to years of U.S. farm work.

<u>Note</u>. Please refer to Table 4 for data pertaining to this section.

- As would be expected, the average age of the farmworkers increased as years of farm work in the U.S. increased (see Figure 6).

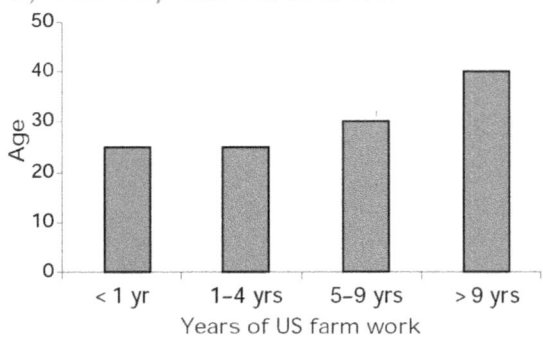

Figure 6. Mean Age of Farmworkers by Years of U.S. Farm Work

- Farmworkers with more than 9 years and with less than 1 year of work on U.S. farms were more likely to be male (85% and 81%, respectively) than those with 1 to 9 years of work on U.S. farms (1 to 4 years, 72%; 5 to 9 years, 68%). The probability that farmworkers were female was highest in the 1 to 4 years (28%) and 5 to 9 years (32%) categories.

- There were fewer immigration-authorized workers in their first year of farm work (11%). Still, a number of workers with more years of farm work were without legal work authorization (see Figure 7).

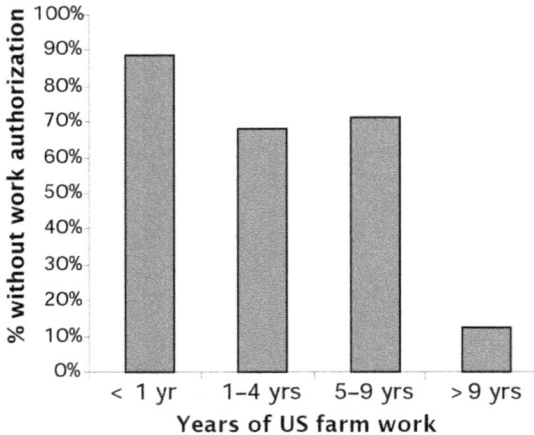

Figure 7. Percentage of Workers Without Legal Authorization to Work in the U.S. by Years of U.S. Farm Work

- The percentage of farmworkers who spoke English only "a little" or "not at all" was 90% for those with less than one year of U.S. farm work. On the other hand, the percentage of workers who spoke English only "a little" or "not at all" was substantially lower (67%) for those with more than 9 years of U.S. farm work.

- Approximately two-thirds of workers with more than 4 years of farm work were settled, meaning they did not move for work.

- More than one-third of farmworkers worked in fruit and nut crops; about one-fourth worked in vegetable crops.

Chapter 4: Results

Table 4. National Agricultural Workers Survey demographic variables by years of work on U.S. farms October 1998—September 1999

	SD	Total	Years working in farm work in the United States			
			<1 yr	1–4 yrs	5–9 yrs	>9 yrs
		Mean (se)[1]	Mean (se)[1]	Mean (se)[1]	Mean (se)[1]	Mean (se)[1]
• Age (mean)	◆	31.4 (0.7)	25.0 (0.6)	25.4 (0.9)	30.0 (0.7)	40.0 (0.6)
		% (se)[1]	% (se)[1]	% (se)[1]	% (se)[1]	% (se)[1]
• Gender male	◆	77.9 (2.5)	80.6 (2.8)	72.0 (6.0)	68.0 (5.0)	84.5 (3.0)
• Not authorized (cannot legally work in the United States)	◆	53.4 (3.6)	88.7 (3.9)	68.1 (6.1)	71.0 (3.3)	12.2 (1.6)
• Speak English 1	◆					
Not at all/a little		77.0 (3.4)	90.3 (3.8)	75.6 (4.7)	81.9 (3.2)	67.0 (3.7)
• Payment	◆					
Hourly		79.8 (3.4)	75.9 (5.8)	84.5 (3.7)	83.5 (2.6)	77.6 (4.1)
Piece rate		15.2 (2.6)	21.0 (5.3)	13.1 (3.4)	13.3 (2.3)	13.6 (2.5)
Combination of hourly and piece rate		2.3 (1.0)	2.4 (1.5)	1.2 (0.6)	1.8 (0.9)	3.1 (1.3)
Salary		2.8 (1.3)	0.8 (0.5)	1.2 (0.7)	1.5 (0.7)	5.8 (3.2)
• Employer	◆					
Farm labor contractor		28.0 (6.2)	36.8 (7.4)	25.4 (6.5)	28.5 (7.1)	23.8 (5.8)
Grower		72.0 (6.2)	63.2 (7.4)	74.6 (6.5)	71.5 (7.1)	76.2 (5.8)
• Migrant type	◆					
Newcomer		22.5 (3.5)	84.4 (4.2)	10.6 (2.6)	0.0 (0.0)	0.0 (0.0)
Follow-the-crop		7.6 (1.5)	0.5 (0.3)	11.1 (3.0)	8.0 (2.2)	9.7 (2.0)
Shuttle		21.2 (2.7)	5.0 (1.7)	30.6 (6.7)	23.1 (4.2)	24.9 (3.2)
Settled		48.7 (3.9)	10.2 (3.1)	47.7 (5.8)	69.0 (5.0)	65.4 (3.8)
• Number of farmworkers employed on farm						
1–10		4.5 (1.0)	2.2 (0.9)	5.2 (1.5)	4.9 (1.2)	5.2 (1.1)
11–50		40.7 (5.9)	42.7 (9.0)	38.0 (6.3)	42.1 (8.1)	40.5 (5.9)
51–150		26.8 (4.2)	30.2 (6.0)	20.8 (4.5)	24.4 (4.7)	29.6 (5.0)
151+		28.1 (6.2)	24.9 (7.4)	36.0 (7.7)	28.7 (6.9)	24.7 (6.2)
• Crop category	◆					
Field crops		15.4 (2.5)	12.9 (3.2)	11.3 (2.7)	11.9 (3.0)	21.1 (4.0)
Fruits and nuts		38.5 (8.0)	41.8 (10.8)	33.1 (8.2)	46.7 (10.0)	36.1 (7.1)
Horticulture		16.3 (4.5)	13.8 (5.2)	25.8 (8.8)	10.6 (3.0)	14.2 (4.1)
Vegetables		25.9 (6.9)	29.0 (10.1)	22.5 (7.5)	25.8 (7.2)	23.3 (5.6)
Miscellaneous/multiple		5.1 (1.4)	2.6 (1.1)	7.3 (3.2)	5.0 (1.6)	5.2 (1.6)

◆ SD=Statistically Different. Rows or groups of rows with an "◆" indicate that differences in prevalence or mean between two or more levels of the stratification variable exist at the p<0.05 level. In cases without an "◆", differences were not statistically significant.

[1] (se) – Standard Error.

Note. Due to rounding, some column totals may not add up to exactly 100 percent.

Chapter 4: Results

Part Two: Data from the Occupational Health Supplement

Starting with Table 5, tables are in the order that the questions appear in the Occupational Health Supplement questionnaire (see Appendix A).

Section One: Participation in Pesticide Safety Training Programs

Tables 5–8 show farmworker participation in pesticide safety training programs for the total population and stratified by years in U.S. farm work, migrant status, number of workers employed on the farm, and crop category.

The NAWS Occupational Health Supplement included questions regarding the WPS to determine whether workers are being adequately trained in pesticide safety. Many of the questions in the Supplement are based on the last 12 months before the interview. However, given that the WPS mandates that all workers must be retrained every 5 years [EPA 1993], a question on whether the worker had received pesticide training in the last 5 years was included.

Pesticide safety training overall

Approximately one-third of farmworkers had not received any pesticide safety training in the last 5 years (See Table 5 and Figure 8).

Pesticide safety training by years working in U.S. farm work, migrant status, number of employees on the farm, and crop category

- Only 41% of workers with less than 1 year of work on U.S. farms received some pesticide safety training during the last 12 months versus 71% for those with 1 to 4 years and 72% for those with five years or more (Table 5).

- As might be expected, newcomers were less likely than follow-the-crop migrants, shuttle migrants, or settled workers to have received some pesticide safety training within the last 12 months (44%, 58%, 76% and 69%, respectively) (Table 6).

- Workers on farms with more than fifty workers were more likely to have received some training within the last 12 months than those on farms with fewer than fifty workers (Table 7).

- The likelihood of not receiving pesticide safety training at any time during the last five years was highest for workers in field crops (52%) when compared to all other crop categories (Table 8) (see Figure 9).

Figure 8. Pesticide Safety Training

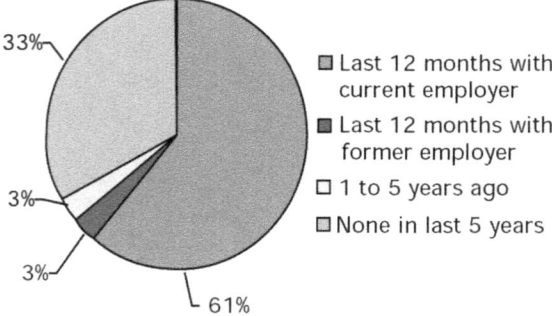

Chapter 4: Results

Figure 9. Percentage of Workers Reporting No Pesticide Safety Training Any Time During the Last Five Years by Crop Category

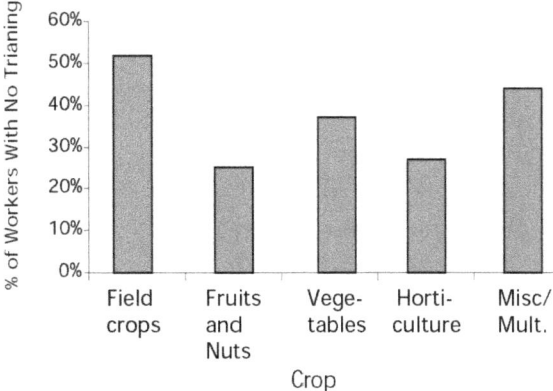

Workers in fruit and nuts (36%) and in vegetables (32%) reported more informal training than workers in other crop categories. Most farmworkers (79%) reported training that lasted an hour or less. Despite the fact that 32% of vegetable workers reported informal training, they were the most likely (31%) to report training that exceeded one hour (Table 8).

Language of pesticide safety training

Eighty-four percent of workers trained said that their training was in Spanish, 12% said it was in English, and 4% said it was bilingual (Spanish and English) (see Table 5 and Figure 11).

Formal versus informal training

For those who were trained, 28% reported that the training consisted of informal training in the fields. Workers with less than 1 year of work on U.S. farms were more likely to have been informally trained; those with more years of work on U.S. farms were more likely to have been formally trained (see Table 5 and Figure 10).

Figure 11. Language of Training

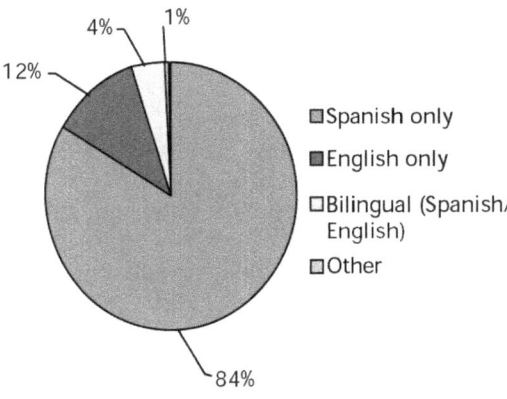

Figure 10. Percentage of Workers Whose Pesticide Safety Training Consisted of Informal Instructions in the Field by Years of U.S. Farm Work

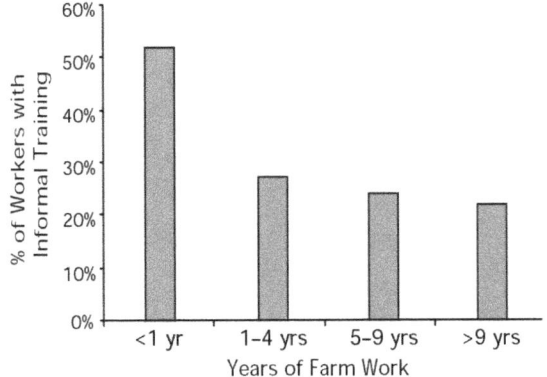

Training was for the most part carried out in the worker's primary language (95%), though 5% were still trained in a language other than their primary (Table 5). Follow-the-crop migrants were the group least likely to receive training in their primary language (91%) compared to shuttle migrants, newcomers, and settled workers (all, 95%) (Table 6). In addition, those on farms with fewer workers were less likely to receive training in their primary language than those on farms with more workers (see Table 7 and Figure 12). Finally, the probability that training was conducted in a farmworker's primary language was lowest in field crops (90%) compared with all other crop categories (Table 8).

Chapter 4: Results

Figure 12. Percent of Workers who Received Pesticide Safety Training in Their Primary Language by Number of Farmworkers Employed on Farm

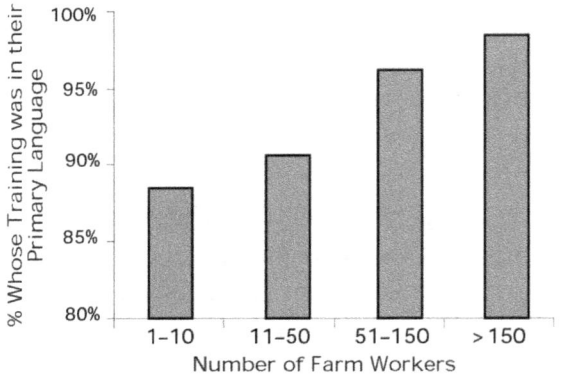

Training based on requirements by EPA's WPS

Eleven percent of workers reported that the training did not cover one or more of the following three topics required by EPA's WPS:

- How soon a worker can enter a field treated with pesticides (2%)
- Injuries or illnesses due to pesticides (6%)
- Where to go or whom to contact for emergency care (10%)

The likelihood that the training covered all three topics increased with more years of U.S. farm work (See Table 5 and Figure 13).

Figure 13. Percentage of Workers Reporting Training that Covered Three WPS Required Topics, by Years of U.S. Farm Work

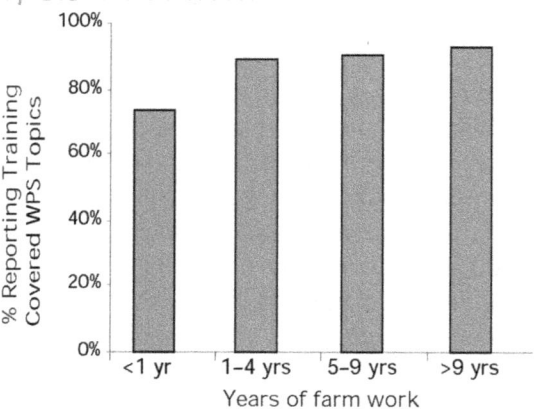

Pesticide safety training certification card

Roughly 70% of farmworkers trained in the last 12 months did not receive a certification card for pesticide safety training (Table 5). Those with less than 5 years of U.S. farm work were the least likely to have received a pesticide training card in the last 12 months compared to those with more years of U.S. farm work (see Table 5 and Figure 14).

Newcomers were also less likely (15%) than follow-the-crop (32%), shuttle (33%), and settled farmworkers (32%) to have received a pesticide training card in the last 12 months (Table 6).

Figure 14. Farmworkers Trained in the Last 12 Months who Received a Certification Card for Pesticide Safety Training, by Years of U.S. Farm Work

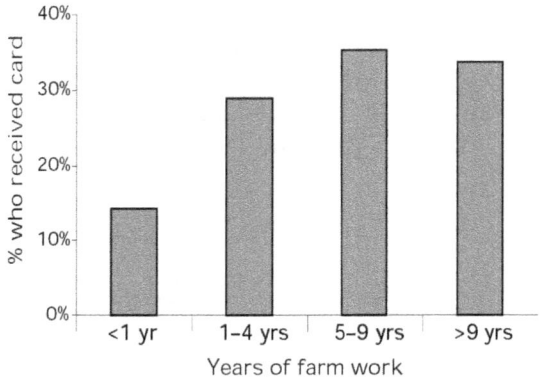

Chapter 4: Results

Table 5. National Agricultural Workers Survey participation in pesticide safety training programs by years of work on U.S. farms, October 1998—September 1999

Pesticide safety training	SD	Total % (se)[1]	Years working in farm work in the United States			
			<1 yr % (se)[1]	1–4 yrs % (se)[1]	5–9 yrs % (se)[1]	>9 yrs % (se)[1]
Did you receive training in the safe use of pesticides?						
• Received some pesticide training, during the last 12 months	♦	64.1 (3.5)	41.1 (4.1)	70.7 (5.4)	71.5 (7.3)	71.5 (4.0)
With your current employer, during the last 12 months	♦	60.9 (3.7)	38.6 (4.3)	66.6 (5.7)	66.4 (6.8)	68.9 (4.3)
With former employer, during the last 12 months		3.3 (0.7)	2.6 (1.1)	4.1 (1.6)	5.1 (1.7)	2.6 (0.6)
• No pesticide training in the last 12 months but did receive training in the last 5 years	♦	2.6 (0.5)	0.3 (0.2)	0.9 (0.4)	1.9 (0.5)	5.6 (1.3)
• No pesticide training any time during the last 5 years	♦	33.3 (3.4)	58.6 (4.1)	28.4 (5.2)	26.6 (7.2)	23.0 (3.8)
• **How was the training delivered?**						
Informal (informal instructions in the field)**	♦	27.9 (6.8)	52.0 (9.7)	26.9 (8.0)	24.0 (8.8)	21.8 (4.8)
Formal (video, audio, written material, class)	♦	71.7 (6.8)	48.0 (9.7)	72.8 (8.0)	75.8 (8.8)	77.7 (4.8)
• **How long was the training or instructions?**						
<½ hour		8.7 (1.7)	10.4 (3.8)	9.4 (2.7)	9.1 (1.9)	7.5 (1.7)
½ hour–1 hour		70.2 (3.5)	70.8 (7.8)	75.1 (4.9)	66.9 (5.4)	68.4 (4.1)
>1 hour		20.8 (3.5)	18.0 (7.4)	15.5 (4.3)	24.0 (5.4)	23.7 (3.7)
• **Who trained or instructed you?*****						
Grower or grower's staff		67.5 (3.7)	70.1 (7.4)	69.8 (4.7)	66.6 (4.1)	65.6 (4.3)
Farm labor contractor or farm labor contractor's staff	♦	13.8 (3.4)	21.8 (8.0)	18.5 (3.7)	11.7 (3.0)	9.1 (2.2)
Government agency	♦	14.5 (1.8)	5.0 (2.7)	9.5 (2.5)	15.1 (2.6)	20.6 (2.9)
Insurance company		5.1 (1.3)	2.3 (1.3)	3.4 (1.1)	8.0 (2.4)	5.8 (1.4)
Other		3.5 (1.1)	2.7 (1.7)	2.6 (1.2)	4.9 (2.3)	3.7 (1.1)
• **In what language(s) was the training or instructions delivered?**						
English only	♦	11.5 (3.0)	7.1 (4.0)	7.8 (3.3)	9.1 (2.9)	16.2 (3.7)
Spanish only	♦	84.0 (3.5)	90.5 (4.3)	89.1 (3.6)	87.6 (3.2)	77.2 (4.3)
Other language		0.7 (0.4)	0.6 (0.4)	0.6 (0.4)	1.2 (0.7)	0.6 (0.5)
Bilingual English/Spanish	♦	4.1 (1.2)	2.2 (1.1)	2.7 (1.1)	2.3 (1.0)	6.5 (2.0)
• Was training in worker's primary language?		94.9 (1.2)	95.2 (2.1)	94.8 (2.1)	93.7 (1.7)	95.4 (1.1)
• **Did the training cover the following topics required by EPA's Worker Protection Standard?**						
How soon you can enter a field treated with pesticides		97.8 (0.6)	99.2 (0.5)	97.7 (0.9)	95.5 (1.6)	98.4 (0.8)
Illness or injuries due to pesticides	♦	94.3 (1.0)	85.8 (4.2)	94.7 (1.6)	94.9 (1.1)	96.8 (1.0)
Where to go or who to contact for emergency medical care	♦	90.1 (1.5)	75.4 (5.8)	90.5 (2.4)	93.7 (1.0)	93.4 (1.4)
• Did the training cover all three topics: Reentry, illness, and emergency care	♦	88.6 (1.5)	73.1 (6.2)	88.9 (2.7)	90.7 (1.6)	92.8 (1.4)
Did you ever receive a certification card for training or instructions in the safe use of pesticides?						
• Received a certification card for pesticide safety training	♦	20.6 (3.0)	5.8 (1.8)	20.9 (6.5)	26.1 (4.5)	27.5 (3.5)
• Farmworkers trained in last 12 months, who received a certification card for pesticide safety training	♦	29.9 (4.7)	14.2 (4.4)	29.1 (8.7)	35.3 (6.1)	33.8 (4.8)

♦ SD=Statistically Different. Rows or groups of rows with an "♦" indicate that differences in prevalence between two or more levels of the stratification variable exist at the p<0.05 level. In cases without an "♦", differences were not statistically significant.

** [1] "Informal" refers to the response "informal instructions in the field." If the respondent also reported "formal" training, they are not included in this category.

*** May report more than one source of training.

[1] (se) – Standard Error.

Note. Due to rounding, some column totals may not add up to exactly 100 percent.

Chapter 4: Results

Table 6. National Agricultural Workers Survey participation in pesticide safety training programs by migrant status, October 1998—September 1999

Pesticide safety training	SD	Total % (se)[1]	Migrant status			
			Newcomer % (se)[1]	Follow-the-crop % (se)[1]	Shuttle % (se)[1]	Settled % (se)[1]
Did you receive training in the safe use of pesticides?						
• Received some pesticide training, during the last 12 months	◆	64.1 (3.5)	44.1 (4.3)	58.0 (5.6)	75.6 (3.7)	69.0 (4.7)
With your current employer, during the last 12 months	◆	60.7 (3.7)	40.9 (4.3)	45.1 (6.9)	73.2 (3.9)	66.6 (4.8)
With former employer, during the last 12 months	◆	3.3 (0.7)	3.2 (1.4)	12.8 (5.0)	2.4 (0.8)	2.4 (0.6)
• No pesticide training in the last 12 months but did receive training in the last 5 years	◆	2.6 (0.5)	0.0 (0.0)	6.7 (2.2)	2.1 (0.7)	3.3 (0.8)
• No pesticide training any time during the last 5 years	◆	33.3 (3.4)	55.9 (4.3)	35.4 (4.7)	22.3 (3.6)	27.7 (4.8)
• **How was the training delivered?**						
Informal (informal instructions in the field)**	◆	27.9 (6.9)	56.6 (10.7)	26.7 (5.4)	30.1 (8.3)	19.2 (5.3)
Formal (video, audio, written material, class)	◆	71.7 (6.8)	43.5 (10.7)	72.7 (5.7)	69.4 (8.3)	80.5 (5.2)
• **How long was the training or instructions?**						
<½ hour		8.7 (1.7)	8.4 (3.6)	15.3 (5.1)	10.6 (2.7)	7.0 (1.4)
½ hour–1 hour		70.2 (3.6)	73.3 (7.9)	68.7 (8.0)	71.4 (5.8)	68.7 (4.4)
>1 hour		20.8 (3.5)	17.5 (7.5)	15.5 (6.8)	18.0 (4.3)	24.0 (4.6)
• **Who trained or instructed you?***						
Grower or grower's staff		67.5 (3.7)	67.6 (6.6)	60.8 (5.9)	74.7 (4.2)	64.7 (4.1)
Farm labor contractor or farm labor contractor's staff	◆	13.8 (3.4)	23.0 (7.5)	18.2 (5.7)	15.6 (3.3)	10.0 (2.4)
Government agency	◆	14.5 (1.8)	5.1 (2.8)	10.2 (2.6)	6.8 (1.9)	21.3 (2.7)
Insurance company		5.1 (1.3)	2.4 (1.2)	6.6 (4.9)	2.7 (1.1)	6.9 (1.8)
Other		3.5 (1.1)	2.9 (1.5)	6.7 (3.0)	3.0 (1.1)	3.6 (1.4)
• **In what language(s) was the training or instructions delivered?**						
English only	◆	11.5 (3.0)	1.2 (1.1)	2.7 (1.8)	4.7 (2.9)	18.2 (4.4)
Spanish only	◆	84.0 (3.5)	97.6 (1.5)	94.9 (2.2)	91.7 (3.4)	75.6 (4.9)
Other language		0.7 (0.4)	0.6 (0.5)	0.0 (0.0)	0.3 (0.3)	1.0 (0.7)
Bilingual English/Spanish		4.1 (1.2)	1.0 (0.6)	2.5 (1.2)	3.3 (1.5)	5.7 (1.7)
• Was training in worker's primary language?		94.9 (1.2)	95.4 (2.0)	90.8 (3.3)	95.3 (2.3)	95.4 (0.9)
• **Did the training cover the following topics required by EPA's Worker Protection Standard?**						
How soon you can enter a field treated with pesticides		97.8 (0.6)	99.4 (0.5)	94.2 (4.0)	98.0 (0.6)	97.9 (0.8)
Illness or injuries due to pesticide	◆	94.3 (1.0)	87.3 (4.2)	95.6 (2.5)	94.2 (2.1)	96.1 (0.8)
Where to go or whom to contact for emergency medical care	◆	90.1 (1.5)	77.6 (5.1)	94.1 (3.8)	88.1 (3.0)	93.8 (1.2)
• Did the training cover all three topics: Reentry, illness, and emergency care	◆	88.6 (1.5)	75.2 (5.4)	92.0 (4.0)	86.9 (3.1)	92.6 (1.2)
Did you ever receive a certification card for training or instructions in the safe use of pesticides?						
• Received a certification card for pesticide safety training	◆	20.6 (3.0)	6.7 (2.0)	23.3 (4.4)	25.6 (7.3)	24.1 (3.5)
• Farmworkers trained in last 12 months, who received a certification card for pesticide safety training		29.9 (4.7)	15.1 (4.8)	32.2 (6.4)	32.7 (8.9)	32.2 (5.1)

◆ SD=Statistically Different. Rows or groups of rows with an "◆" indicate that differences in prevalence between two or more levels of the stratification variable exist at the p<0.05 level. In cases without an "◆", differences were not statistically significant.

** [1] "Informal" refers to the response "informal instructions in the field." If the respondent also reported "formal" training, they are not included in this category.

*** May report more than one source of training.

[1] (se) – Standard Error.   Note. Due to rounding, some column totals may not add up to exactly 100 percent.

Chapter 4: Results

Table 7. National Agricultural Workers Survey participation in pesticide safety training programs by number of farmworkers employed on farm, October 1998–September 1999

Pesticide safety training	SD	Total % (se)[1]	Number of farmworkers employed on farm			
			1–10 % (se)[1]	11–50 % (se)[1]	51–150 % (se)[1]	>150 % (se)[1]
Did you receive training in the safe use of pesticides?						
• Received some pesticide training during the last 12 months	◆	64.1 (3.5)	54.4 (4.4)	49.4 (4.5)	65.4 (6.4)	85.7 (2.5)
With your current employer, during the last 12 months	◆	60.7 (3.7)	46.3 (4.5)	45.3 (4.6)	62.6 (6.7)	83.7 (2.9)
With former employer, during the last 12 months		3.3 (0.7)	8.2 (1.8)	4.1 (1.2)	2.8 (1.1)	2.0 (1.1)
• No pesticide training in the last 12 months but did receive training in the last 5 years	◆	2.6 (0.5)	3.7 (0.9)	3.9 (0.9)	2.3 (0.9)	0.9 (0.3)
• No pesticide training any time during the last 5 years	◆	33.3 (3.4)	41.9 (4.3)	46.8 (4.7)	32.3 (6.2)	13.4 (2.3)
• **How was the training delivered?**						
Informal (informal instructions in the field)**		27.9 (6.8)	24.4 (4.9)	23.5 (4.0)	29.3 (12.4)	31.3 (9.4)
Formal (video, audio, written material, class)		71.7 (6.8)	72.6 (5.2)	76.3 (4.1)	70.5 (12.3)	68.5 (9.4)
• **How long was the training or instructions?**						
<½ hour		8.7 (1.7)	14.6 (3.7)	12.5 (2.5)	9.6 (2.0)	4.0 (2.0)
½ hour–1 hour		70.2 (3.5)	48.9 (7.2)	66.3 (4.2)	70.5 (4.6)	75.6 (4.9)
>1 hour		20.8 (3.5)	36.5 (8.9)	20.8 (3.4)	19.1 (4.9)	20.4 (5.4)
• **Who trained or instructed you?***						
Grower or grower's staff		67.5 (3.7)	49.0 (7.0)	65.5 (4.6)	75.3 (5.4)	65.4 (6.0)
Farm labor contractor or farm labor contractor's staff		13.8 (3.4)	16.1 (7.7)	8.8 (2.4)	13.4 (5.0)	18.3 (5.3)
Government agency		14.5 (1.8)	25.5 (6.1)	17.8 (3.6)	10.6 (2.6)	13.3 (2.4)
Insurance company		5.1 (1.3)	1.9 (1.3)	4.7 (1.9)	3.7 (1.7)	6.9 (2.1)
Other	◆	3.5 (1.1)	8.8 (2.9)	6.3 (1.96)	2.7 (1.5)	1.1 (0.6)
• **In what language(s) was the training or instructions delivered?**						
English only	◆	11.5 (3.0)	35.5 (8.2)	26.8 (6.3)	4.5 (1.6)	0.5 (0.3)
Spanish only	◆	84.0 (3.5)	53.2 (6.5)	67.1 (6.1)	90.2 (2.6)	97.7 (1.1)
Other language		0.7 (0.4)	0.0 (0.0)	0.5 (0.3)	1.3 (0.8)	0.5 (0.5)
Bilingual English/Spanish		4.1 (1.2)	11.4 (3.9)	5.6 (1.8)	4.5 (1.6)	1.8 (1.0)
• Was training in worker's primary language?		94.9 (1.2)	88.5 (6.9)	90.6 (2.0)	96.2 (1.2)	98.5 (1.0)
• **Did the training cover the following topics required by EPA's Worker Protection Standard?**						
How soon you can enter a field treated with pesticides	◆	97.8 (0.6)	96.6 (1.2)	94.4 (1.4)	99.8 (0.2)	99.5 (0.2)
Illness or injuries due to pesticides		94.3 (1.0)	93.2 (2.7)	94.2 (1.1)	95.7 (1.2)	93.5 (1.9)
Where to go or who to contact for emergency medical care		90.1 (1.5)	89.8 (4.1)	90.6 (1.8)	92.3 (2.7)	88.0 (2.1)
• Did the training cover all three topics: Reentry, illness, and emergency care		88.6 (1.5)	86.2 (5.0)	88.1 (2.1)	91.5 (2.6)	87.0 (2.2)
Did you ever receive a certification card for training or instructions in the safe use of pesticides?						
• Received a certification card for pesticide safety training		20.6 (3.1)	22.1 (5.2)	22.6 (3.9)	14.5 (2.5)	23.2 (6.8)
• Farmworkers trained in last 12 months, who received a certification card for pesticide safety training		29.9 (4.7)	37.1 (7.4)	40.5 (5.9)	21.6 (4.0)	26.4 (7.8)

◆ SD=Statistically Different. Rows or groups of rows with an "◆" indicate that differences in prevalence between two or more levels of the stratification variable exist at the p<0.05 level. In cases without an "◆", differences were not statistically significant.

** 1 "Informal" refers to the response "informal instructions in the field." If the respondent also reported "formal" training, they are not included in this category.

*** May report more than one source of training.

[1] (se) – Standard Error.

Note. Due to rounding, some column totals may not add up to exactly 100 percent.

Chapter 4: Results

Table 8. National Agricultural Workers Survey participation in pesticide safety training programs by crop category, October 1998—September 1999

Pesticide safety training	SD	Total % (se)[1]	Crop categories				
			Field crops % (se)[1]	Fruit and Nuts % (se)[1]	Vegetables % (se)[1]	Horticulture % (se)[1]	Misc/mult % (se)[1]
Did you receive training in the safe use of pesticides?							
• Received some pesticide training during the last 12 months	◆	64.1 (3.5)	43.0 (6.5)	73.4 (4.4)	60.2 (8.0)	71.1 (6.4)	53.4 (12.4)
With your current employer, during the last 12 months	◆	60.7 (3.7)	38.5 (6.7)	69.2 (4.8)	57.5 (8.6)	70.0 (6.5)	50.1 (11.9)
With former employer, during the last 12 months		3.3 (0.7)	4.6 (2.5)	4.2 (1.2)	2.8 (1.0)	1.1 (0.6)	3.3 (2.3)
• No pesticide training in the last 12 months but did receive training in the last 5 years		2.6 (0.5)	5.2 (1.7)	1.8 (0.7)	3.0 (0.9)	1.7 (0.7)	2.3 (1.3)
• No pesticide training any time during the last 5 years	◆	33.3 (3.4)	51.8 (7.0)	24.8 (4.3)	36.8 (8.1)	27.2 (6.3)	44.3 (12.9)
• How was the training delivered?							
Informal (informal instructions in the field)**		27.9 (6.8)	19.8 (5.0)	36.3 (13.1)	32.0 (8.3)	11.0 (3.7)	12.2 (4.0)
Formal (video, audio, written material, class)		71.7 (6.8)	80.2 (5.0)	63.4 (13.1)	67.2 (7.9)	88.9 (3.7)	87.8 (4.0)
• How long was the training or instructions?							
<½ hour		8.7 (1.7)	10.1 (3.3)	7.8 (2.2)	6.2 (3.0)	13.4 (2.6)	7.8 (3.4)
½ hour–1 hour		70.2 (3.5)	71.5 (3.9)	74.3 (5.4)	62.3 (5.2)	66.5 (7.4)	83.0 (5.8)
>1 hour		20.8 (3.5)	18.4 (2.9)	17.5 (4.9)	30.8 (5.5)	20.0 (6.3)	9.2 (4.7)
• Who trained or instructed you?***							
Grower or grower's staff	◆	67.5 (3.7)	54.0 (8.0)	64.2 (6.3)	64.6 (3.9)	86.3 (3.2)	73.9 (9.6)
Farm labor contractor or farm labor contractor's staff		13.8 (3.4)	10.2 (3.3)	17.0 (6.7)	17.0 (3.6)	7.1 (3.8)	1.8 (1.0)
Government agency	◆	14.5 (1.8)	29.9 (7.4)	12.5 (2.1)	13.2 (2.9)	8.8 (3.4)	25.8 (10.3)
Insurance company		5.1 (1.3)	6.3 (2.3)	7.7 (2.5)	2.8 (1.7)	1.7 (0.9)	2.3 (1.3)
Other		3.5 (1.1)	5.8 (2.2)	3.0 (1.8)	5.6 (2.3)	1.3 (0.5)	0.0 (0.0)
• In what language(s) was the training or instructions delivered?							
English only	◆	11.5 (3.0)	44.1 (11.6)	2.9 (0.9)	5.1 (2.3)	20.6 (8.1)	11.9 (6.8)
Spanish only	◆	84.0 (3.5)	51.3 (11.1)	92.7 (1.9)	89.7 (4.6)	75.2 (8.7)	86.0 (6.6)
Other language		0.7 (0.4)	0.1 (0.1)	1.3 (0.9)	0.1 (0.1)	0.7 (0.6)	0.0 (0.0)
Bilingual English/Spanish		4.1 (1.2)	4.6 (2.3)	3.9 (1.3)	5.2 (2.9)	3.6 (1.6)	2.1 (1.4)
• Was training in worker's primary language?		94.9 (1.2)	90.1 (5.9)	95.8 (1.0)	95.5 (1.7)	94.7 (2.2)	96.2 (2.9)
• Did the training cover the following topics required by EPA's Worker Protection Standard?							
How soon you can enter a field treated with pesticides	◆	97.8 (0.6)	93.4 (2.9)	98.3 (0.5)	99.2 (0.4)	98.3 (1.0)	95.1 (3.2)
Illness or injuries due to pesticides		94.3 (1.0)	92.0 (2.4)	94.5 (1.3)	93.7 (2.8)	96.3 (1.6)	94.3 (3.1)
Where to go or who to contact for emergency medical care		90.1 (1.5)	86.7 (3.8)	89.7 (2.5)	90.9 (3.8)	90.5 (3.4)	97.1 (1.8)
• Did the training cover all three topics: Reentry, illness and emergency care		88.6 (1.5)	85.0 (3.5)	87.6 (2.1)	90.8 (3.8)	89.0 (3.8)	93.4 (3.4)
Did you ever receive a certification card for training or instructions in the safe use of pesticides?							
• Received a certification card for pesticide safety training		20.6 (3.0)	22.8 (5.2)	20.8 (4.3)	19.7 (4.2)	18.7 (9.5)	22.3 (7.7)
• Farmworkers trained in last 12 months, who received a certification card for pesticide safety training		29.9 (4.7)	44.0 (7.5)	27.0 (6.1)	30.3 (7.4)	25.7 92.8)	40.6 11.4)

◆ SD=Statistically Different. Rows or groups of rows with an "◆" indicate that differences in prevalence between two or more levels of the stratification variable exist at the p<0.05 level. In cases without an "◆", differences were not statistically significant.

** "Informal" refers to the response "informal instructions in the field." If the respondent also reported "formal" training, they are not included in this category.

*** May report more than one source of training.

[1] (se) – Standard Error.

Note. Due to rounding, some column totals may not add up to exactly 100 percent.

Chapter 4: Results

Section Two: Personal Protective Equipment (PPE) Worn by Pesticide Loaders, Mixers, or Applicators

Tables 9 through 12 show the percentage of farmworkers who loaded, mixed and applied pesticides in the United States in the last 12 months and which types of PPE were used the last time they worked in these jobs. Data is shown for the total population and is also stratified by years working in U.S. farm work, migrant status, number of employees on the farm, and crop category.

Percentage of farmworkers who loaded, mixed, or applied pesticides in the United States in the last 12 months

Overall, 11% of farmworkers reported loading, mixing or applying pesticides (Table 9).

- The probability of performing one of these tasks increased with more years of U.S. farm work (Table 9).

- The likelihood of performing one of these tasks was greater for those employed on farms with fewer workers than for those employed on farms with more workers (see Table 11 and Figure 15).

- Workers in field crops and workers in miscellaneous/multiple crops were more likely to have performed one of these tasks (21% and 17%, respectively) than workers in other crop categories (Fruit and nut 9%; vegetables 5%; horticulture 13%) (Table 12).

Figure 15. Percentage of Farmworkers who Loaded, Mixed or Applied Pesticides in the U.S. in the Last 12 Months, by number of Farmworkers Employed on Farm

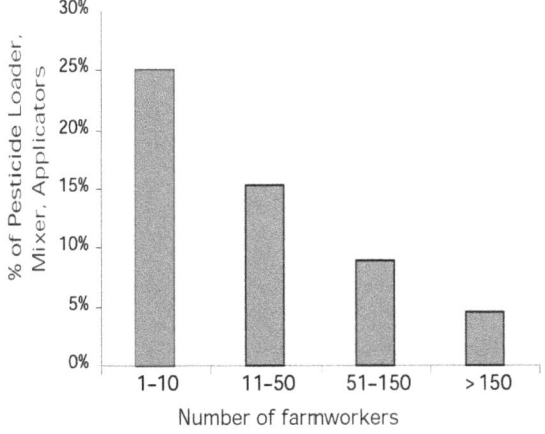

Personal protective equipment worn by pesticide loaders, mixers, or applicators during the last pesticide-related task performed in the last 12 months

Prevalence of farmworkers using personal protective equipment generally was higher with more years of work on U.S. farms (see Table 9 and Figure 16). In addition, those working on the smallest farms (1 to 10 farmworkers) reported the lowest percentage of respirator (46%) and goggle use (58%), while those working on the largest farms (> 150 farmworkers) reported the highest percentage (73% and 74%, respectively) (Table 11).

Figure 16. Use of Personal Protective Equipment, by Years of U.S. Farm work

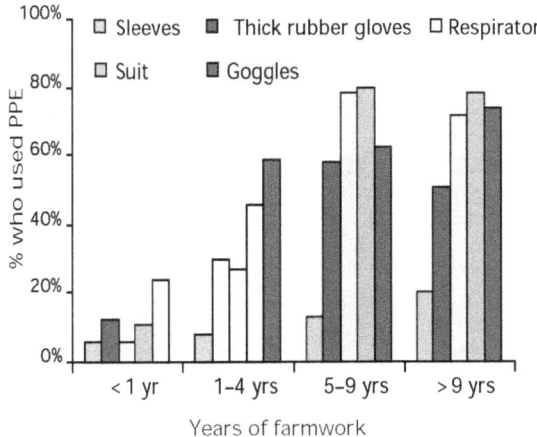

Chapter 4: Results

Table 9. National Agricultural Workers Survey personal protective equipment worn by pesticide loaders, mixers, or applicators during the last pesticide-related task performed** in the last 12 months, by years of work on U.S. farms, October 1998—September 1999

Pesticide loaders, mixers, or applicators	SD	Total % (se)[1]	Years working in farm work in the United States			
			<1 yr % (se)[1]	1–4 yrs % (se)[1]	5–9 yrs % (se)[1]	>9 yrs % (se)[1]
Have you loaded, mixed, or applied pesticides in the United States in the last 12 months?	◆	11.0 (1.9)	3.3 (1.7)	6.8 (1.5)	10.6 (3.0)	18.9 (3.0)
The last time you loaded, mixed, or applied pesticides did you wear:***						
• Any type of ppe (thick rubber gloves, sleeves, suit, respirator, goggles)	◆	84.0 (4.1)	26.2 (15.8)	73.4 (8.8)	90.3 (4.4)	91.4 (2.4)
• Gloves	◆					
Thick rubber		46.2 (4.3)	11.9 (8.1)	30.1 (7.0)	58.4 (6.9)	51.3 (4.7)
Thin rubber		23.5 (2.6)	14.4 (8.4)	40.7 (9.2)	14.7 (5.1)	22.5 (3.3)
Cloth		17.7 (3.0)	18.6 (11.8)	13.0 (3.8)	15.6 (5.1)	19.0 (4.3)
None		12.5 (3.6)	55.0 (13.8)	16.2 (7.2)	11.4 (6.1)	7.2 (2.2)
• Sleeves		15.7 (4.3)	5.6 (5.2)	7.6 (3.7)	12.7 (7.8)	19.7 (5.0)
• Suit	◆	69.1 (4.7)	10.8 (7.2)	46.1 (9.3)	79.8 (9.0)	78.2 (3.3)
• Respirator****	◆	61.4 (4.1)	5.6 (5.2)	27.4 (6.2)	77.5 (7.7)	71.6 (3.4)
• Goggles	◆	66.2 (5.2)	23.8 (15.5)	58.5 (10.8)	62.0 (8.9)	74.1 (4.3)

◆ SD=Statistically Different. Rows or groups of rows with an "◆" indicate that differences in prevalence between two or more levels of the stratification variable exist at the p<0.05 level. In cases without an "◆", differences were not statistically significant.
** If "Last Time" involved the use of a toxicity class III pesticide, there may be no requirement to use some of this PPE (suit, respirator, goggles).
*** Respondents could list more than one type of personal protective equipment.
**** Refers to a respirator other than a bandana or paper mask, including NIOSH certified filtering face piece particulate dust masks such as the N95.
[1] (se) – Standard Error.

Table 10. National Agricultural Workers Survey personal protective equipment worn by pesticide loaders, mixers, or applicators during the last pesticide-related task performed** in the last 12 months, by migrant status, October 1998—September 1999

Pesticide loaders, mixers, or applicators	SD	Total % (se)[1]	Migrant status			
			Newcomer % (se)[1]	Follow-the-crop % (se)[1]	Shuttle % (se)[1]	Settled % (se)[1]
Have you loaded, mixed, or applied pesticides in the United States in the last 12 months?	◆	11.0 (1.9)	3.0 (1.7)	5.5 (1.8)	7.2 (1.7)	17.1 (3.1)
The last time you loaded, mixed, or applied pesticides did you wear:***						
• Any type of ppe (thick rubber gloves, sleeves, suit, respirator, goggles)	◆	84.0 (4.1)	15.2 (10.2)	89.4 (6.9)	86.7 (6.6)	88.5 (2.6)
• Gloves	◆					
Thick rubber		46.2 (4.3)	11.9 (8.7)	52.1 (16.9)	38.6 (7.0)	50.9 (4.5)
Thin rubber		23.5 (2.6)	23.0 (9.7)	22.3 (11.5)	25.7 (9.9)	22.3 (3.2)
Cloth		17.7 (3.0)	1.3 (1.4)	20.3 (15.7)	18.9 (7.9)	18.6 (3.9)
None		12.5 (3.6)	63.8 (12.2)	5.3 (4.0)	16.8 (6.1)	8.2 (2.1)
• Sleeves		15.7 (4.3)	6.7 (6.4)	0.9 (1.0)	10.0 (4.3)	18.5 (5.3)
• Suit	◆	69.1 (4.7)	13.6 (9.5)	84.6 (9.8)	67.1 (11.7)	73.7 (3.7)
• Respirator****	◆	61.4 (4.1)	7.4 (6.6)	59.7 (12.4)	65.9 (10.5)	65.3 (3.2)
• Goggles	◆	66.2 (5.2)	10.1 (7.9)	54.2 (17.0)	65.5 (8.9)	70.7 (4.4)

◆ SD=Statistically Different. Rows or groups of rows with an "◆" indicate that differences in prevalence between two or more levels of the stratification variable exist at the p<0.05 level. In cases without an "◆", differences were not statistically significant.
** If "Last Time" involved the use of a toxicity class III pesticide, there may be no requirement to use some of this PPE (suit, respirator, goggles).
*** Respondents could list more than one type of personal protective equipment.
**** Refers to a respirator other than a bandana or paper mask, including NIOSH certified filtering face piece particulate dust masks such as the N95.
[1] (se) – Standard Error.

Chapter 4: Results

Table 11. National Agricultural Workers Survey personal protective equipment worn by pesticide loaders, mixers, or applicators during the last pesticide-related task performed** in the last 12 months, by number of farmworkers employed on farm, October 1998—September 1999

Pesticide loaders, mixers, or applicators	SD	Total % (se)[1]	Number of farmworkers employed on farm			
			1–10 % (se)[1]	11–50 % (se)[1]	51–150 % (se)[1]	>150 % (se)[1]
Have you loaded, mixed, or applied pesticides in the United States in the last 12 months?	◆	11.0 (1.9)	25.1 (4.6)	15.4 (2.8)	8.9 (2.6)	4.5 (1.5)
The last time you loaded, mixed, or applied pesticides did you wear:***						
• Any type of ppe (thick rubber gloves, sleeves, suit, respirator, goggles)		84.0 (4.1)	87.3 (5.9)	84.9 (3.9)	79.0 (10.6)	86.2 (11.1)
• Gloves						
Thick rubber		46.2 (4.3)	61.5 (7.8)	39.1 (5.0)	53.5 (7.8)	54.5 (15.9)
Thin rubber		23.5 (2.6)	11.3 (3.8)	26.0 (4.3)	18.5 (4.3)	31.6 (14.8)
Cloth		17.7 (2.9)	16.3 (3.7)	23.4 (3.8)	7.2 (3.0)	10.4 (7.4)
None		12.5 (3.6)	10.9 (5.2)	11.5 (4.0)	20.8 (10.2)	3.5 (3.6)
• Sleeves		15.7 (4.3)	14.5 (5.5)	14.1 (4.0)	27.5 (13.1)	3.1 (2.4)
• Suit		69.1 (4.7)	68.9 (8.1)	68.9 (4.7)	66.5 (11.3)	75.2 (19.0)
• Respirator****		61.4 (4.1)	45.8 (7.9)	62.2 (4.8)	60.8 (8.8)	72.6 (18.7)
• Goggles		66.2 (5.2)	57.8 (8.9)	65.0 (6.0)	69.0 (12.1)	74.4 (18.7)

◆ SD=Statistically Different. Rows or groups of rows with an "◆" indicate that differences in prevalence between two or more levels of the stratification variable exist at the p<0.05 level. In cases without an "◆", differences were not statistically significant.
** If "Last Time" involved the use of a toxicity class III pesticide, there may be no requirement to use some of this PPE (suit, respirator, goggles).
*** Respondents could list more than one type of personal protective equipment.
**** Refers to a respirator other than a bandana or paper mask, including NIOSH certified filtering face piece particulate dust masks such as the N95.
[1] (se) – Standard Error.

Table 12. National Agricultural Workers Survey personal protective equipment worn by pesticide loaders, mixers, or applicators during the last pesticide-related task performed** in the last 12 months, by crop category, October 1998—September 1999

Pesticide loaders, mixers, or applicators	SD	Total % (se)[1]	Crop categories				
			Field crops % (se)[1]	Fruit and Nuts % (se)[1]	Vegetables % (se)[1]	Horticulture % (se)[1]	Misc/Mult % (se)[1]
Have you loaded, mixed, or applied pesticides in the United States in the last 12 months?	◆	11.0 (1.9)	20.8 (5.4)	9.2 (3.0)	5.1 (1.7)	13.1 (2.8)	17.4 (5.9)
The last time you loaded, mixed, or applied pesticides did you wear:***							
• Any type of ppe (thick rubber gloves, sleeves, suit, respirator, goggles)		84.0 (4.1)	80.4 (6.9)	86.3 (6.0)	83.3 (9.3)	83.4 (8.4)	90.1 (5.7)
• Gloves	◆						
Thick rubber		46.2 (4.3)	46.4 (8.4)	52.1 (6.8)	45.1 (9.9)	43.2 (9.4)	31.2 (9.7)
Thin rubber		23.5 (2.6)	16.1 (3.5)	25.1 (4.9)	24.7 (7.4)	25.7 (7.8)	37.4 (13.5)
Cloth		17.7 (2.9)	22.5 (5.6)	15.1 (5.1)	13.6 (7.3)	15.8 (4.5)	21.1 (9.1)
None		12.5 (3.6)	15.0 (6.7)	7.7 (3.7)	16.6 (7.5)	15.3 (10.7)	10.3 (6.6)
• Sleeves		15.7 (4.3)	13.5 (6.7)	24.4 (10.2)	7.6 (3.8)	9.5 (2.0)	15.9 (10.3)
• Suit		69.1 (4.7)	61.6 (6.1)	75.9 (8.3)	75.1 (9.2)	61.6 (11.0)	78.8 (9.8)
• Respirator****		61.4 (4.1)	57.6 (6.5)	63.7 (7.3)	62.8 (10.1)	61.5 (9.8)	63.9 (14.0)
• Goggles		66.2 (5.2)	57.2 (8.9)	70.1 (9.0)	66.4 (11.5)	70.7 (13.4)	72.0 (9.3)

◆ SD=Statistically Different. Rows or groups of rows with an "◆" indicate that differences in prevalence between two or more levels of the stratification variable exist at the p<0.05 level. In cases without an "◆", differences were not statistically significant.
** If "Last Time" involved the use of a toxicity class III pesticide, there may be no requirement to use some of this PPE (suit, respirator, goggles).
*** Respondents could list more than one type of personal protective equipment.
**** Refers to a respirator other than a bandana or paper mask, including NIOSH certified filtering face piece particulate dust masks such as the N95.
[1] (se) – Standard Error.

Chapter 4: Results

Section Three: Availability of Drinking Water, Toilets, and Hand Washing Facilities

Availability of sanitary facilities

Tables 13 to 16 show the distribution of sanitary facilities that were available to farmworkers at their workplace for the total population and stratified by years working in U.S. farm work, migrant status, number of employees on the farm, and crop category.

- Twenty-two percent of workers did not have water and disposable cups available (Table 13).

- Fourteen percent did not have toilets and toilet paper (Table 13).

- Twenty-three percent did not have water and hand washing supplies available to them on a daily basis (Table 13).

The availability of supplies including water and disposable cups, toilet and sufficient toilet paper, and hand washing water, soap, and single use towels, declined with the number of farmworkers employed on the farm (see Table 15 and Figure 17).

More workers in field crops and miscellaneous/multiple crops consistently lacked these supplies than workers in other crop categories (Table 16).

In most cases, farmworkers with more than ten years of work on U.S. farms lacked these supplies to a greater extent than workers with fewer years of work on U.S. farms. Still, farmworkers with less than 1 year of U.S. farm work reported a shortage of hand washing water and supplies (25%) more often than those with more years of farm work (Table 13).

Follow-the-crop workers did without water and disposable cups (30%) as well as hand washing water and supplies (35%) more frequently than newcomers, shuttle, or settled workers; however, settled workers were more likely to go without toilet and sufficient toilet paper (16%) (Table 14).

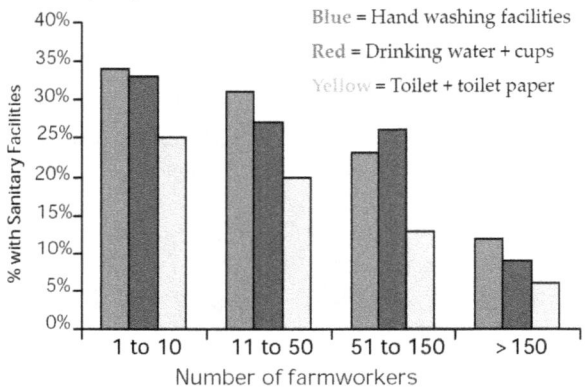

Figure 17. Availability of Drinking Water, Toilets, and Hand washing Facilities by Number of Farmworkers Employed on Farm

*Table 13. National Agricultural Workers Survey
availability of drinking water, toilets, and hand washing facilities,
by years of work on U.S. farms,
October 1998—September 1999*

			Years working in farm work in the United States			
		Total	<1 yr	1–4 yrs	5–9 yrs	>9 yrs
Does your current employer provide (Every day):	SD	% (se)[1]	% (se)[1]	% (se)[1]	% (se)[1]	% (se)[1]
Drinking water						
No water		6.2 (1.5)	4.4 (1.7)	3.8 (1.4)	5.4 (1.8)	9.5 (3.0)
Lacked water or disposable cups		21.9 (4.1)	19.0 (7.1)	20.0 (4.3)	20.1 (4.7)	25.8 (4.6)
Toilet						
No toilet		9.6 (2.1)	8.4 (2.8)	9.7 (3.8)	6.1 (1.6)	12.0 (3.5)
Lacked toilet or sufficient toilet paper		14.3 (2.4)	12.6 (3.1)	12.9 (3.9)	9.7 (2.3)	18.6 (3.9)
Hand washing water						
No hand washing water		10.0 (2.2)	6.7 (2.1)	10.9 (3.9)	8.7 (2.0)	12.2 (3.4)
Lacked hand washing water, soap, or single use towels		23.4 (3.0)	25.3 (4.9)	21.9 (4.4)	20.2 (3.6)	24.8 (4.4)

♦ SD=Statistically Different. Rows or groups of rows with an "♦" indicate that differences in prevalence between two or more levels of the stratification variable exist at the p<0.05 level. In cases without an "♦", differences were not statistically significant.

[1] (se) – Standard Error.

Note. Due to rounding, some column totals may not add up to exactly 100 percent.

*Table 14. National Agricultural Workers Survey
availability of drinking water, toilets, and hand washing facilities,
by migrant status,
October 1998—September 1999*

			Migrant status			
		Total	Newcomer	Follow-the-crop	Shuttle	Settled
Does your current employer provide (Every day):	SD	% (se)[1]	% (se)[1]	% (se)[1]	% (se)[1]	% (se)[1]
Drinking water						
No water		6.2 (1.5)	4.5 (2.0)	5.0 (2.0)	4.9 (1.2)	7.9 (2.5)
Lacked water or disposable cups		21.9 (4.1)	18.2 (7.1)	27.8 (5.6)	24.3 (4.5)	21.7 (4.4)
Toilet						
No toilet		9.6 (2.1)	7.5 (2.7)	7.8 (3.4)	7.2 (1.8)	12.0 (3.5)
Lacked toilet or sufficient toilet paper		14.4 (2.4)	12.1 (3.1)	16.1 (4.4)	11.4 (2.3)	16.5 (3.7)
Hand washing water						
No hand washing water		10.0 (2.2)	7.1 (2.6)	13.3 (4.1)	8.6 (2.0)	11.5 (3.5)
Lacked hand washing water, soap, or single use towels		23.4 (3.0)	25.5 (5.0)	34.9 (4.8)	22.3 (3.4)	21.3 (4.0)

♦ SD=Statistically Different. Rows or groups of rows with an "♦" indicate that differences in prevalence between two or more levels of the stratification variable exist at the p<0.05 level. In cases without an "♦", differences were not statistically significant.

[1] (se) – Standard Error.

Note. Due to rounding, some column totals may not add up to exactly 100 percent.

Chapter 4: Results

Table 15. National Agricultural Workers Survey availability of drinking water, toilets, and hand washing facilities, by number of farmworkers employed on farm, October 1998—September 1999

Does your current employer provide (Every day):	SD	Total % (se)[1]	Number of farmworkers employed on farm			
			1–10 % (se)[1]	11–50 % (se)[1]	51–150 % (se)[1]	>150 % (se)[1]
Drinking water						
No water		6.2 (1.5)	10.8 (3.1)	8.6 (1.9)	8.2 (4.3)	0.2 (0.2)
Lacked water or disposable cups	♦	21.9 (4.1)	32.6 (4.9)	26.9 (4.4)	26.3 (6.7)	8.7 (4.6)
Toilet						
No Toilet	♦	9.6 (2.1)	21.0 (5.7)	15.1 (3.2)	7.6 (4.3)	1.7 (1.5)
Lacked toilet or sufficient toilet paper	♦	14.3 (2.4)	25.3 (5.6)	20.2 (3.4)	12.6 (5.8)	5.7 (1.5)
Hand washing water						
No hand washing water		10.0 (2.2)	16.9 (5.3)	15.6 (3.5)	9.2 (4.3)	1.6 (1.4)
Lacked hand washing water, soap, or single use towels	♦	23.4 (3.0)	33.7 (5.4)	30.7 (3.8)	22.7 (6.8)	12.0 (3.6)

♦ SD=Statistically Different. Rows or groups of rows with an "♦" indicate that differences in prevalence between two or more levels of the stratification variable exist at the p<0.05 level. In cases without an "♦", differences were not statistically significant.

[1] (se) – Standard Error.

Note. Due to rounding, some column totals may not add up to exactly 100 percent.

Table 16. National Agricultural Workers Survey availability of drinking water, toilets, and hand washing facilities, by crop category, October 1998—September 1999

Does your current employer provide (Every day):	SD	Total % (se)[1]	Crop categories				
			Field crops % (se)[1]	Fruit and Nuts % (se)[1]	Vegetables % (se)[1]	Horticulture % (se)[1]	Misc/Mult % (se)[1]
Drinking water							
No water	♦	6.2 (1.5)	18.7 (6.1)	2.9 (1.2)	3.8 (1.7)	4.7 (2.2)	10.7 (3.8)
Lacked water or disposable cups		21.9 (4.1)	36.7 (6.2)	11.4 (4.1)	26.6 (9.9)	23.1 (7.3)	29.9 (9.7)
Toilet							
No Toilet	♦	9.6 (2.1)	32.8 (6.5)	2.6 (1.0)	5.9 (2.3)	3.5 (1.8)	30.1 (15.1)
Lacked toilet or sufficient toilet paper	♦	14.3 (2.4)	42.8 (6.7)	5.4 (1.3)	12.8 (2.8)	4.3 (1.9)	35.4 (14.3)
Hand washing water							
No hand washing water	♦	10.0 (2.2)	30.5 (7.5)	3.1 (1.0)	8.3 (2.8)	2.9 (1.5)	31.7 (15.9)
Lacked hand washing water, soap, or single use towels	♦	23.4 (3.0)	49.9 (7.5)	12.9 (2.7)	27.4 (5.3)	10.6 (2.8)	45.0 (12.7)

♦ SD=Statistically Different. Rows or groups of rows with an "♦" indicate that differences in prevalence between two or more levels of the stratification variable exist at the p<0.05 level. In cases without an "♦", differences were not statistically significant.

[1] (se) – Standard Error.

Note. Due to rounding, some column totals may not add up to exactly 100 percent.

Chapter 4: Results

Section Four: Health Conditions and Symptoms

Tables 17 through 20 contain estimated 12-month prevalences of musculoskeletal pain or discomfort, respiratory symptoms, dermatitis, and gastrointestinal problems for the total population. The data is also stratified by years of U.S. farm work, migrant status, number of farmworkers employed on farm, and crop category. In most cases, farmworkers who have worked more years (Table 17), settled farmworkers (Table 18), workers on farms with fewer workers (Table 19), and farmworkers working in miscellaneous/multiple crops (Table 20) reported higher prevalences of these conditions.

Musculoskeletal pain or discomfort

Fifteen percent of farmworkers reported musculoskeletal pain or discomfort every day for a week or more in one or more of the following body parts:

- Back (6%)
- Shoulder/ neck and upper extremities (5%)
- Lower extremities (4%)

Farmworkers with 10 years of work or more on U.S. farms had the highest prevalence of pain for all 3 body parts (back, 9%; shoulder/ neck/ upper extremities, 6%; lower extremities, 5%) compared with workers who had fewer years of work on U.S. farms (Table 17). This may be due to age's association with years of farm work. When controlling for age, years of farm work is only significantly related to musculoskeletal pain in any body part for those farmworkers older than age 36 (data not shown). Overall musculoskeletal pain or discomfort was more frequent for farmworkers on the smallest farms (<11 workers) (27%) than it was for those on farms with 11 or more workers. (see Table 19 and Figure 18).

Dermatitis

Dermatitis was reported by nearly 7% of the hired farmworkers overall (Table 17). It was less frequent among farmworkers employed on farms with more than 150 workers (5%) than on farms with fewer workers (see Table 19 and Figure 18). In all crop categories, with the exception of those in miscellaneous and multiple crops, farmworkers were more likely to have dermatitis on their hands and arms as opposed to other body parts. Compared with workers in other crop categories, workers in miscellaneous and multiple crops had the highest prevalence of dermatitis overall (8%), as well as on the torso and legs (4%). The prevalence of dermatitis in workers in fruits and nuts was slightly less than 8%, with more cases of dermatitis affecting the face (4%) compared with workers in other crop categories (Table 20).

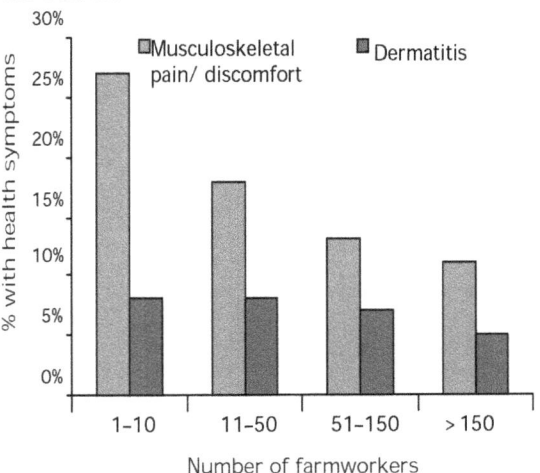

Figure 18. Musculoskeletal Pain/ Discomfort and Dermatitis Symptoms by Number of Farmworkers Employed on Farm

Chapter 4: Results

Respiratory symptoms

"Runny stuffy nose or watery itchy eyes" was the most common respiratory symptom reported (13.8%). Wheezing or whistling in chest was reported by 3.1% of farmworkers, and having coughed or brought up phlegm on most days for at least 3 months was reported by almost 2% of workers (Table 17). Prevalences of all respiratory symptoms were higher for settled farmworkers compared with newcomers, follow-the-crop, and shuttle migrant workers (Table 18). Each of these respiratory symptoms was more pervasive for workers on farms with <11 workers (runny stuffy nose or watery itchy eyes, 25%; wheezing or whistling in the chest, 8%; coughed or brought up phlegm, 3%). (see Table 19 and Figure 19).

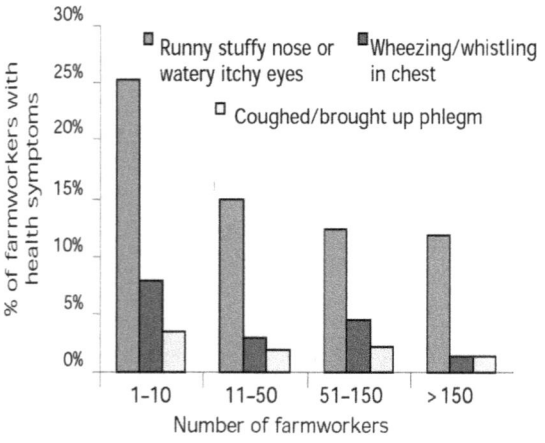

Figure 19. Respiratory Symptoms by Number of Farmworkers Employed on Farm

Gastrointestinal problems

Diarrhea lasting more than 3 days was uncommon, reported by approximately 3% of all farmworkers (Table 17).

Chapter 4: Results

Table 17. National Agricultural Workers Survey estimated 12-month prevalence of health conditions and symptoms, by years of work on U.S. farms October 1998—September 1999

Health condition or symptom	SD	Total % (se)[1]	Years working in farm work in the United States			
			<1 yr % (se)[1]	1–4 yrs % (se)[1]	5–9 yrs % (se)[1]	>9 yrs % (se)[1]
Musculoskeletal pain or discomfort						
• In the last 12 months, have you had any pain or discomfort?	◆	14.9 (1.9)	10.7 (2.4)	14.8 (2.4)	11.7 (2.1)	19.3 (3.0)
• Reported pain or discomfort every day for a week or more in the last 12 months that affected the following areas:**						
Back	◆	6.4 (1.1)	4.3 (1.4)	5.5 (1.3)	6.2 (1.2)	8.5 (1.8)
Shoulder/neck and upper extremities		4.7 (0.7)	4.6 (1.1)	3.2 (0.8)	4.0 (1.3)	6.1 (1.1)
Lower extremities		3.6 (1.0)	3.5 (1.8)	2.7 (0.8)	2.9 (1.3)	4.6 (1.3)
Respiratory symptoms						
• Have you had wheezing or whistling in your chest at any time in the last 12 months?	◆	3.1 (0.5)	1.5 (0.7)	2.2 (0.6)	2.9 (0.8)	4.7 (0.8)
• Have you had episodes of runny stuffy nose or watery itchy eyes?	◆	13.8 (2.3)	5.1 (1.5)	13.3 (2.5)	18.1 (6.9)	17.9 (3.5)
• Have you coughed or brought up phlegm on most days for at least 3 months?	◆	1.8 (0.4)	0.7 (0.5)	0.8 (0.3)	2.8 (1.1)	2.9 (0.6)
Dermatitis						
• In the last 12 months, have you had any skin problem such as redness, inflammation, discoloration, or rash?		6.9 (0.8)	6.7 (1.5)	6.7 (1.1)	7.8 (1.8)	6.8 (1.1)
• Reported dermatitis in the last 12 months that affected the following areas:**						
Hands and arms		4.7 (0.6)	3.5 (1.4)	5.2 (1.0)	5.9 (1.4)	4.6 (0.9)
Face		2.1 (0.5)	2.9 (1.2)	1.3 (0.6)	2.2 (1.1)	2.1 (0.7)
Other, including torso and legs		1.9 (0.3)	2.0 (0.8)	1.3 (0.5)	2.4 (1.2)	1.9 (0.4)
Gastrointestinal problem						
• Diarrhea that lasted more than 3 days		2.5 (0.4)	2.7 (0.8)	2.1 (0.6)	1.9 (0.9)	2.9 (0.9)

◆ SD=Statistically Different. Rows or groups of rows with an "◆" indicate that differences in prevalence between two or more levels of the stratification variable exist at the p<0.05 level. In cases without an "◆", differences were not statistically significant.

** Some individuals reported more than one area.

[1] (se) – Standard Error.

Note. Due to rounding, some column totals may not add up to exactly 100 percent.

Chapter 4: Results

Table 18. National Agricultural Workers Survey
estimated 12-month prevalence of health conditions and symptoms,
by migrant status,
October 1998—September 1999

Health condition or symptom	SD	Total % (se)[1]	Migrant status			
			Newcomer % (se)[1]	Follow-the-crop % (se)[1]	Shuttle % (se)[1]	Settled % (se)[1]
Musculoskeletal pain or discomfort						
• In the last 12 months, have you had any pain or discomfort?		14.9 (1.9)	11.3 (2.6)	16.7 (3.5)	15.4 (2.8)	16.2 (2.6)
• Reported pain or discomfort every day for a week or more in the last 12 months that affected the following areas:**						
Back		6.4 (1.1)	5.1 (1.9)	4.1 (1.8)	4.9 (1.4)	8.1 (1.3)
Shoulder/neck and upper extremities		4.7 (0.7)	4.8 (1.1)	4.6 (1.3)	5.5 (1.3)	4.4 (0.8)
Lower extremities		3.6 (1.0)	2.7 (1.6)	2.9 (1.3)	4.2 (1.4)	3.9 (1.1)
Respiratory symptoms						
• Have you had wheezing or whistling in your chest at any time in the last 12 months?	♦	3.1 (0.5)	1.3 (0.7)	1.3 (0.6)	1.6 (0.5)	4.8 (0.8)
• Have you had episodes of runny stuffy nose or watery itchy eyes?	♦	13.8 (2.3)	4.8 (1.3)	10.6 (2.1)	13.4 (3.2)	18.8 (3.8)
• Have you coughed or brought up phlegm on most days for at least 3 months?		1.8 (0.4)	0.7 (0.5)	2.1 (0.7)	1.2 (0.5)	2.6 (0.7)
Dermatitis						
• In the last 12 months, have you had any skin problem such as redness, inflammation, discoloration, or rash?		6.9 (0.8)	6.7 (1.6)	5.7 (1.2)	7.4 (1.7)	7.1 (1.1)
• Reported dermatitis in the last 12 months that affected the following areas:**						
Hands and arms		4.7 (0.6)	4.1 (1.7)	5.0 (1.2)	4.7 (1.1)	4.9 (0.9)
Face		2.1 (0.5)	3.0 (1.3)	0.8 (0.4)	1.7 (0.8)	2.1 (0.7)
Other, including torso and legs		1.9 (0.3)	1.5 (0.7)	1.1 (0.4)	2.0 (0.7)	2.1 (0.5)
Gastrointestinal problems						
• Diarrhea that lasted more than 3 days		2.5 (0.4)	2.2 (0.9)	1.7 (0.8)	2.4 (0.5)	2.7 (0.8)

♦ SD=Statistically Different. Rows or groups of rows with an "♦" indicate that differences in prevalence between two or more levels of the stratification variable exist at the $p<0.05$ level. In cases without an "♦", differences were not statistically significant.

** Some individuals reported more than one area.

[1] (se) – Standard Error.

Note. Due to rounding, some column totals may not add up to exactly 100 percent.

Chapter 4: Results

Table 19. National Agricultural Workers Survey estimated 12-month prevalence of health conditions and symptoms, by number of farmworkers employed on farm, October 1998–September 1999

Health condition or symptom	SD	Total %(se)[1]	Number of farmworkers employed on farm			
			1–10 %(se)[1]	11–50 %(se)[1]	51–150 %(se)[1]	>150 %(se)[1]
Musculoskeletal pain or discomfort						
• In the last 12 months have you had any pain or discomfort?		14.9 (1.9)	26.5 (3.9)	17.9 (2.5)	12.9 (2.3)	10.6 (3.8)
• Reported pain or discomfort every day for a week or more in the last 12 months that affected the following areas:**						
Back		6.4 (1.1)	9.8 (2.4)	6.3 (1.1)	6.9 (1.7)	5.6 (2.2)
Shoulder/neck and upper extremities		4.7 (0.7)	8.5 (2.9)	4.6 (0.8)	4.4 (1.0)	4.5 (1.5)
Lower extremities		3.6 (1.0)	3.0 (1.3)	2.7 (0.8)	4.7 (1.6)	4.1 (2.0)
Respiratory symptoms						
• Have you had wheezing or whistling in your chest at any time in the last 12 months?	◆	3.1 (0.5)	7.8 (2.1)	2.9 (0.6)	4.4 (1.0)	1.3 (0.4)
• Have you had episodes of runny stuffy nose or watery itchy eyes?		13.8 (2.3)	25.3 (5.1)	14.9 (3.3)	12.3 (2.3)	11.8 (3.9)
• Have you coughed or brought up phlegm on most days for at least 3 months?		1.8 (0.4)	3.4 (0.9)	1.9 (0.5)	2.1 (0.9)	1.2 (0.7)
Dermatitis						
• In the last 12 months, have you had any skin problem such as redness, inflammation, discoloration, or rash?		6.9 (0.8)	7.8 (2.6)	7.9 (1.3)	7.1 (1.3)	5.3 (1.5)
• Reported dermatitis in the last 12 months that affected the following areas:**						
Hands and arms		4.7 (0.6)	5.1 (1.7)	5.5 (1.0)	5.4 (1.5)	2.9 (1.1)
Face		2.1 (0.5)	2.0 (0.7)	1.6 (0.5)	2.6 (1.1)	2.4 (1.1)
Other, including torso and legs		1.9 (0.3)	2.8 (1.7)	1.8 (0.5)	2.2 (0.6)	1.5 (0.5)
Gastrointestinal problem						
• Diarrhea that lasted more than 3 days		2.5 (0.4)	3.0 (1.1)	2.7 (0.7)	3.5 (0.9)	1.2 (0.4)

◆ SD=Statistically Different. Rows or groups of rows with an "◆" indicate that differences in prevalence between two or more levels of the stratification variable exist at the p<0.05 level. In cases without an "◆", differences were not statistically significant.

** Some individuals reported more than one area.

[1] (se) – Standard Error.

Note. Due to rounding, some column totals may not add up to exactly 100 percent.

Chapter 4: Results

Table 20. National Agricultural Workers Survey estimated 12-month prevalence of health conditions and symptoms, by crop category, October 1998—September 1999

Health condition or symptom	SD	Total %(se)[1]	Crop categories				
			Field crops %(se)[1]	Fruit and nuts %(se)[1]	Vegetables %(se)[1]	Horiculture %(se)[1]	Misc/mult %(se)[1]
Musculoskeletal pain or discomfort							
• In the last 12 months have you had any pain or discomfort?		14.9 (1.9)	16.0 (2.7)	15.5 (4.1)	15.2 (2.0)	10.5 (1.4)	19.9 (6.7)
• Reported pain or discomfort every day for a week or more in the last 12 months that affected the following areas:**							
Back		6.4 (1.1)	5.8 (1.5)	7.2 (2.4)	6.5 (1.4)	4.9 (1.2)	7.1 (3.3)
Shoulder/neck and upper extremities		4.7 (0.7)	4.9 (1.1)	4.9 (1.3)	5.3 (0.9)	2.4 (0.6)	7.2 (3.6)
Lower extremities		3.6 (1.0)	2.7 (1.2)	4.1 (2.3)	4.0 (1.1)	1.5 (0.3)	8.0 (3.6)
Respiratory symptoms							
• Have you had wheezing or whistling in your chest at any time in the last 12 months?		3.1 (0.5)	3.6 (1.1)	2.7 (0.7)	2.6 (0.7)	3.6 (1.1)	4.8 (2.5)
• Have you had episodes of runny stuffy nose or watery itchy eyes?		13.8 (2.3)	11.1 (2.4)	15.4 (4.6)	10.7 (1.4)	18.0 (6.8)	12.0 (5.4)
• Have you coughed or brought up phlegm on most days for at least 3 months?		1.8 (0.4)	3.0 (1.2)	1.5 (0.4)	1.8 (0.8)	1.9 (0.5)	1.3 (0.8)
Dermatitis							
• In the last 12 months, have you had any skin problem such as redness, inflammation, discoloration, or rash?		6.9 (0.8)	6.8 (1.8)	7.8 (1.7)	5.7 (1.1)	6.7 (1.3)	8.0 (4.4)
• Reported dermatitis in the last 12 months that affected the following areas:**							
Hands and arms		4.7 (0.6)	5.5 (1.7)	4.2 (1.0)	4.8 (0.7)	5.3 (1.4)	3.4 (2.0)
Face							
Other, including torso and legs		1.9 (0.3)	1.9 (0.8)	2.3 (0.4)	1.3 (0.3)	0.9 (0.4)	4.2 (2.7)
Gastrointestinal problem							
• Diarrhea that lasted more than 3 days		2.5 (0.4)	2.0 (0.9)	3.3 (0.9)	2.2 (0.6)	2.3 (1.0)	0.6 (0.5)

♦ SD=Statistically Different. Rows or groups of rows with an "♦" indicate that differences in prevalence between two or more levels of the stratification variable exist at the p<0.05 level. In cases without an "♦", differences were not statistically significant.

** Some individuals reported more than one area.

[1] (se) – Standard Error.

Note. Due to rounding, some column totals may not add up to exactly 100 percent.

Chapter 4: Results

Section Five: Smoking and Alcohol Use

Tables 21 through 24 present data on the lifetime and current cigarette use by farmworkers and prevalence of alcohol use among farmworkers for the total population and by years in U.S. farm work, migrant status, number of employees on the farm, and crop category. The tables also include the amount of alcohol they reported consuming, on average, for those who drink.

Smoking

One in four farmworkers were current smokers. Farmworkers who had worked 5 years or more were more likely to smoke than those with fewer years of farm work and were also the most likely to have smoked in the past and quit. Smoking was less prevalent among farmworkers with 1 to 4 years of farm work (20%) than those with more or fewer years of farm work. In addition, farmworkers who had worked less than one year on U.S. farms were less likely to have been former smokers (1%) than those with more years of work on U.S. farms (Table 21). Regarding migrant status, newcomers (20%) were the least likely to be current or former smokers (Table 22). The likelihood that farmworkers were current or former smokers was lower for those working on large farms (>150 workers) (20%) than it was for those on farms with fewer workers (Table 23). Of all crop categories, smoking was most common amongst farmworkers in field crops (33%) (See Table 24 and Figure 20).

Alcohol use

Half of farmworkers reported drinking alcohol in the last month (Table 21). Among those who reported drinking alcohol, they consumed an average of about 38 alcoholic drinks per month. On average, the farmworkers consumed alcohol approximately 10 days during the month. The percentage of farmworkers who consumed alcohol in the last month was greater for those with more than 9 years of farm work in the United States (57%) than for those with fewer years of work on U.S. farms (Table 21). Those who worked with miscellaneous or multiple crops did not drink as often as workers on other crops, but consumed more on the occasions they drank and, therefore, averaged more drinks total over the month (49 drinks) (see Table 24 and Figure 20).

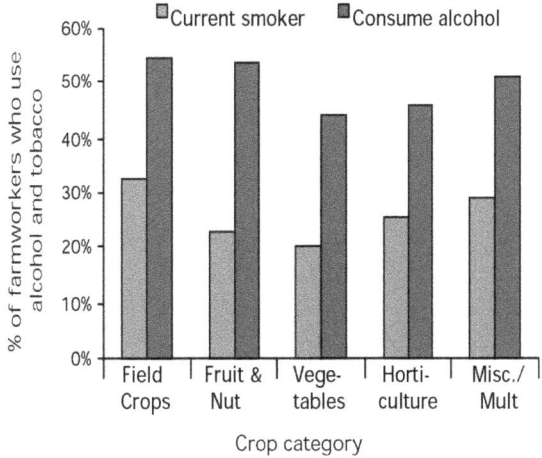

Figure 20. Percentage of farmworkers who are current smokers and percentage who consumed alcohol in the last month, by crop category

Chapter 4: Results

Table 21. National Agricultural Workers Survey smoking and alcohol use, by years of work on U.S. farms, October 1998—September 1999

Alcohol and tobacco use	SD	Total	Years working in farm work in the United States			
			<1 yr	1–4 yrs	5–9 yrs	>9 yrs
Smoking status		% (se)[1]	% (se)[1]	% (se)[1]	% (se)[1]	% (se)[1]
Current smoker	♦	24.6 (1.5)	23.4 (2.5)	20.1 (2.5)	24.7 (3.0)	28.3 (2.6)
Former smoker (have smoked in past, but not in last 12 months)		3.5 (0.7)	1.3 (1.1)	2.8 (1.1)	2.6 (0.9)	5.7 (1.2)
Alcohol consumption	♦	% (se)[1]	% (se)[1]	% (se)[1]	% (se)[1]	% (se)[1]
Percent who consumed alcohol in last month		50.3 (2.1)	49.5 (4.3)	43.8 (4.5)	45.4 (3.5)	57.4 (2.6)
		Mean (se)[1]	Mean (se)[1]	Mean (se)[1]	Mean (se)[1]	Mean (se)[1]
Average days per month that alcohol was consumed**		9.6 (0.7)	9.8 (0.8)	8.7 (0.7)	9.6 (1.2)	9.8 (0.8)
Average number of drinks consumed on occasions they consumed alcohol**		4.5 (0.3)	4.2 (0.6)	4.3 (0.3)	4.8 (0.4)	4.6 (0.3)
Average number of alcoholic drinks consumed per month**		38.5 (1.9)	37.4 (4.2)	34.3 (2.7)	41.8 (3.4)	40.2 (2.5)

♦ SD=Statistically Different. Rows or groups of rows with an "♦" indicate that differences in prevalence or mean between two or more levels of the stratification variable exist at the p<0.05 level. In cases without an "♦", differences were not statistically significant.
** Of those who consumed alcohol.
[1] (se) – Standard Error.

Table 22. National Agricultural Workers Survey smoking and alcohol use, by migrant status, October 1998—September 1999

Alcohol and tobacco use	SD	Total	Migrant status			
			Newcomer	Follow-the-crop	Shuttle	Settled
Smoking status		% (se)[1]	% (se)[1]	% (se)[1]	% (se)[1]	% (se)[1]
Current smoker		24.6 (1.5)	19.8 (1.8)	27.1 (3.4)	28.0 (3.9)	24.8 (2.5)
Former smoker (have smoked in past, but not in last 12 months)		3.5 (0.7)	2.4 (1.3)	6.0 (2.4)	3.3 (0.9)	3.6 (0.8)
Alcohol consumption		% (se)[1]	% (se)[1]	% (se)[1]	% (se)[1]	% (se)[1]
Percent who consumed alcohol in last month		50.2 (2.1)	49.9 (4.9)	49.3 (5.1)	51.0 (5.3)	50.2 (2.9)
		Mean (se)[1]	Mean (se)[1]	Mean (se)[1]	Mean (se)[1]	Mean (se)[1]
Average days per month that alcohol was consumed**	♦	9.6 (0.7)	10.1 (1.0)	9.1 (0.7)	11.3 (1.2)	8.7 (0.5)
Average number of drinks consumed on occasions they consumed alcohol**		4.5 (0.3)	4.0 (0.6)	5.4 (0.3)	4.8 (0.6)	4.4 (0.2)
Average number of alcoholic drinks consumed per month**		38.5 (1.9)	35.1 (4.3)	46.5 (3.7)	44.9 (3.2)	36.0 (2.4)

♦ SD=Statistically Different. Rows or groups of rows with an "♦" indicate that differences in prevalence or mean between two or more levels of the stratification variable exist at the p<0.05 level. In cases without an "♦", differences were not statistically significant.
** Of those who consumed alcohol.
[1] (se) – Standard Error.

Chapter 4: Results

Table 23. National Agricultural Workers Survey smoking and alcohol use, by number of farmworkers employed on farm, October 1998–September 1999

Alcohol and tobacco use	SD	Total	Number of farmworkers employed on farm			
			1–10	11–50	51–150	>150
Smoking status		% (se)[1]	% (se)[1]	% (se)[1]	% (se)[1]	% (se)[1]
Current smoker		24.6 (1.5)	26.6 (4.0)	27.5 (2.9)	25.3 (2.8)	19.6 (2.2)
Former smoker (have smoked in past, but not in last 12 months)		3.5 (0.7)	6.1 (1.5)	3.4 (0.7)	4.5 (1.6)	2.1 (1.2)
Alcohol consumption		% (se)[1]	% (se)[1]	% (se)[1]	% (se)[1]	% (se)[1]
Percent who consumed alcohol in last month		50.3 (2.1)	54.8 (4.9)	48.1 (3.3)	56.6 (3.5)	46.6 (4.1)
		Mean (se)[1]	Mean (se)[1]	Mean (se)[1]	Mean (se)[1]	Mean (se)[1]
Average days per month that alcohol was consumed**	◆	9.6 (0.7)	8.3 (0.8)	8.1 (0.4)	10.1 (1.0)	11.3 (1.1)
Average number of drinks consumed on occasions they consumed alcohol**	◆	4.5 (0.3)	5.4 (0.7)	5.3 (0.3)	4.3 (0.6)	3.2 (0.3)
Average number of alcoholic drinks consumed per month**		38.5 (1.9)	41.6 (3.4)	42.4 (3.1)	37.7 (3.8)	33.3 (3.7)

◆ SD=Statistically Different. Rows or groups of rows with an "◆" indicate that differences in prevalence or mean between two or more levels of the stratification variable exist at the p<0.05 level. In cases without an "◆", differences were not statistically significant.
** Of those who consumed alcohol.
[1] (se) – Standard Error.

Table 24. National Agricultural Workers Survey smoking and alcohol use, by crop category, October 1998–September 1999

Alcohol and tobacco use	SD	Total	Crop categories				
			Field crops	Fruit and Nuts	Vege-tables	Horti-culture	Misc/Mult
Smoking status		% (se)[1]	% (se)[1]	% (se)[1]	% (se)[1]	% (se)[1]	% (se)[1]
Current smoker		24.6 (1.5)	32.5 (5.2)	23.1 (1.9)	20.0 (1.7)	26.4 (5.4)	29.4 (8.3)
Former smoker (have smoked in past, but not in last 12 months)		3.5 (0.7)	3.8 (1.1)	4.6 (1.4)	2.0 (0.9)	3.1 (1.1)	2.1 (1.2)
Alcohol consumption		% (se)[1]	% (se)[1]	% (se)[1]	% (se)[1]	% (se)[1]	% (se)[1]
Percent who consumed alcohol in last month		50.3 (2.1)	55.2 (4.2)	54.2 (2.8)	43.9 (3.0)	45.7 (6.4)	51.1 (12.6)
		Mean (se)[1]	Mean (se)[1]	Mean (se)[1]	Mean (se)[1]	Mean (se)[1]	Mean (se)[1]
Average days per month that alcohol was consumed**		9.6 (0.7)	9.3 (0.6)	10.2 (1.2)	10.4 (1.0)	7.6 (1.0)	8.3 (0.9)
Average number of drinks consumed on occasions they consumed alcohol**		4.5 (0.3)	5.1 (0.3)	4.1 (0.6)	4.7 (0.7)	4.1 (0.2)	5.7 (0.8)
Average number of alcoholic drinks consumed per month**		38.5 (1.9)	44.4 (2.5)	35.3 (3.5)	42.7 (4.0)	31.2 (4.3)	49.0 (10.0)

◆ SD=Statistically Different. Rows or groups of rows with an "◆" indicate that differences in prevalence or mean between two or more levels of the stratification variable exist at the p<0.05 level. In cases without an "◆", differences were not statistically significant.
** Of those who consumed alcohol.
[1] (se) – Standard Error.

Chapter 4: Results

Section Six: Access to and Quality of Health Care

Included in the survey were questions on access to health care, payment for that care, and whether farm workers found it easy or difficult to obtain health care. Information about dental visits was also collected to ascertain the availability of dental care. Tables 25 to 28 contain data from this section of the questionnaire for the entire population as well as by years of work on U.S. farms, migrant status, number of farmworkers employed on farm, and crop category.

Use of health services in the last two years

Approximately 2 out of 3 farmworkers (64%) had not used any health care services in the United States in the last 2 years. Logically, workers with less than 1 year of farm work experience in the United States (17%) used health care services less in the last 2 years than those with more years of work on U.S. farms (Table 25). Likewise, newcomers (11%) reported less use of health care services in the past 2 years than follow-the-crop, shuttle, and settled workers (see Table 26 and Figure 21).

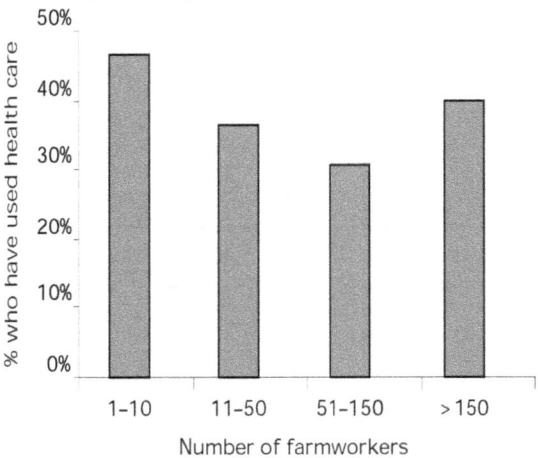

Figure 22. Use of health care services in the United States in the Last 2 years, by number of farmworkers employed on farm

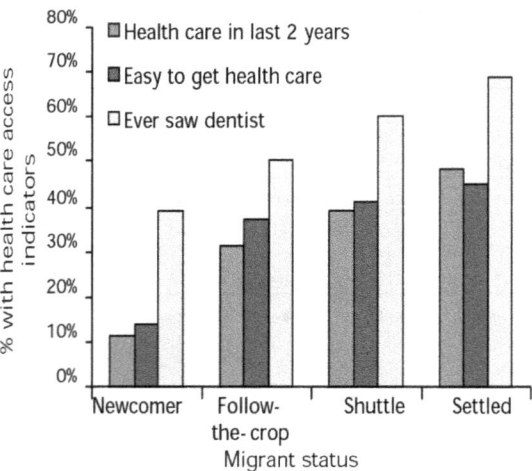

Figure 21. Health care access by migrant status

Workers on the smallest farms (1 to 10 workers) had higher use of health care services in the last two years (46%) than those on farms with more workers (see Table 27 and Figure 22).

For the various crop categories, "miscellaneous/multiple" had the highest use of health care services in the last 2 years (58%) followed by horticulture (46%), vegetables (34%), fruits and nuts (33%), and field crops (31%) (Table 28).

When asked when they had last seen a dentist, 41% of farmworkers replied that they had never seen a dentist either in the United States or elsewhere. Farmworkers with less than 1 year of farm work (56%) were more likely to reply that they had never seen a dentist than those with more years of farm work. Furthermore, the probability that farmworkers had ever seen a dentist diminished with fewer years of work on U.S. farms (see Table 25).

When looking at migrant status, newcomers (39%) were the group the least likely to have ever seen a dentist followed by, follow-the-crop (50%), shuttle (60%), and settled (69%) farmworkers (see Table 26 and Figure 21).

Farmworkers on farms with 11 to 150 workers were less likely to have ever seen a dentist than those employed on farms with more or with fewer workers (Table 27). Finally, those employed in field crops (49%) were the least likely to have ever seen a dentist (Table 28).

Health service and method of payment by whether it was related to farm work

It was a concern that many of the workers may not completely understand the difference between employer-provided health plans and workers compensation. As a result, employer-provided health plans and workers compensation were combined into one category for method of payment. Other method of payment categories included "paid self" and "other."

Four percent of farmworkers reported that their most recent health care visit was related to their farm work job. For those who sought health care related to farm work, 23% paid for the health care out of their pocket, while 59% reported that workers compensation or an employer provided health plan paid for their health care. Farmworkers who had worked for 10 years or more on U.S. farms had a smaller percentage of work-related health care visits (50%) paid for with employer-provided health plans or workers compensation than those with fewer years of work on U.S. farms (Table 25). Farmworkers in fruits and nuts (8%) were the least likely to pay for work-related health care themselves and workers in vegetables (49%) were the most likely compared with those in other crop categories (Table 28).

For visits not related to farm work, 60% of farmworkers paid for health care themselves, however, the percentage of those who paid with employer-provided health plans or workers compensation did increase with more years of farm work (see Table 25 and Figure 23). Workers in miscellaneous/multiple crops (87%) paid for health care not related

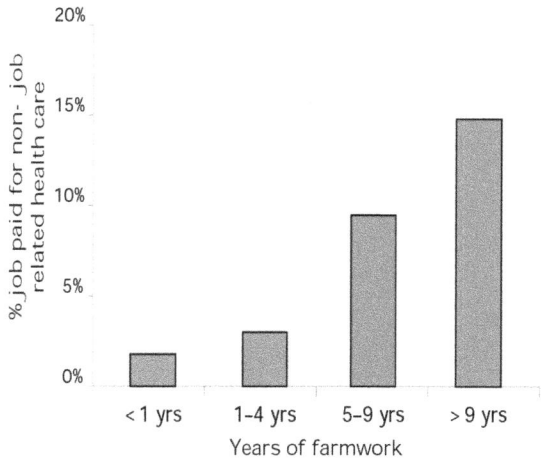

Figure 23. Employment Paid Health Care not Related to Farm Work, by years of farm work

to work themselves more frequently than workers in other crop categories (see Table 28).

Accessibility of health care services in the United States

More than half of farmworkers (51%) reported that it was difficult to get health care services in the United States (Table 25). Follow-the-crop workers (57%) reported the most difficulty in getting health care services in the United States, followed by newcomers (54%), shuttle (51%), and settled workers (48%) (Table 26). The likelihood of reporting difficulty in getting health care services in the United States was greater for workers on farms with more than 50 workers than for those on farms with 50 or fewer workers (Table 27). Workers employed in the fruit and nut category and vegetable category were more likely to report difficulty in getting health care services in the United States (55% and 54%, respectively) than those employed in other crop categories (Table 28).

Chapter 4: Results

Farmworkers who had worked fewer than 5 years in the United States were more likely to answer that they did not know if it was easy or difficult to get health care services in the United States than those who had worked more years (Table 25). As might be expected, newcomers reported that they did not know if it was easy or difficult to get health care services in the U.S. (31%) more often than follow-the-crop (6%), shuttle (8%), and settled workers (7%) (Table 26). In addition, farmworkers on farms with more than 50 workers were more likely to say that they did not know if it is easy or difficult to get health care services in the U.S. than those on farms with fewer workers (Table 27).

Table 25. National Agricultural Workers Survey access to and quality of health care, by years of work on U.S. farms, October 1998—September 1999

			Years working in farm work in the United States			
		Total	<1 yr	1–4 yrs	5–9 yrs	>9 yrs
Health care	SD	% (se)[1]	% (se)[1]	% (se)[1]	% (se)[1]	% (se)[1]
Have you used health care services in the United States in last 2 years?	◆	36.4 (2.3)	17.4 (3.3)	41.3 (5.1)	37.9 (4.0)	44.8 (3.4)
• The last time you used health care services was it:						
Related to your farm work job?	◆	4.2 (0.8)	1.1 (0.6)	5.3 (1.6)	5.3 (1.8)	5.2 (1.0)
Not related to any job?		30.9 (2.2)	15.8 (2.8)	35.1 (5.6)	32.3 (3.7)	37.4 (3.1)
Did not use health care services in last 2 years	◆	63.6 (2.3)	82.6 (3.3)	58.7 (5.1)	62.1 (4.0)	55.2 (3.4)
For those who used health care services in the last 2 years, how did you pay?						
• For those with a problem related to the farm work job?						
Paid self		23.1 (6.0)	5.7 (6.3)	29.4 (12.8)	16.9 (6.6)	24.2 (9.4)
Employer provided health plan or Workers Compensation		59.3 (3.7)	58.5 (27.8)	62.5 (14.0)	77.9 (8.6)	49.9 (10.9)
Other		17.0 (5.1)	35.8 (27.8)	8.1 (4.2)	5.3 (3.4)	25.9 (7.7)
• For those with a problem NOT related to the farm work job?	◆					
Paid self		59.8 (7.8)	64.9 (7.9)	58.3 (6.2)	66.0 (4.8)	55.8 (4.0)
Employer provided health plan or Workers Compensation		9.2 (2.4)	1.7 (1.3)	3.0 (1.5)	9.5 (4.5)	14.8 (3.3)
Other		31.5 (3.0)	33.4 (8.0)	38.8 (5.9)	24.5 (2.8)	29.5 (3.0)
• Is it easy or difficult to get the health care services you need in the United States?	◆					
Difficult		50.6 (2.9)	49.2 (5.4)	53.0 (3.6)	56.2 (3.9)	47.4 (4.0)
Easy		36.8 (3.3)	20.6 (4.7)	36.3 (4.2)	39.5 (3.9)	46.5 (3.8)
Don't know		12.7 (2.1)	30.2 (5.4)	10.7 (2.2)	4.3 (0.8)	6.1 (1.8)
• When was the last time you saw a dentist (In the United States or elsewhere)?						
Never		40.9 (2.7)	55.8 (4.5)	39.7 (4.8)	36.9 (4.7)	33.6 (4.0)

◆ SD=Statistically Different. Rows or groups of rows with an "◆" indicate that differences in prevalence between two or more levels of the stratification variable exist at the p<0.05 level. In cases without an "◆", differences were not statistically significant.

[1] (se) – Standard Error.

Note. Due to rounding, some column totals may not add up to exactly 100 percent.

Chapter 4: Results

Table 26. National Agricultural Workers Survey access to and quality of health care, by migrant status, October 1998—September 1999

Health care	SD	Total % (se)[1]	Migrant status			
			Newcomer % (se)[1]	Follow-the-crop % (se)[1]	Shuttle % (se)[1]	Settled % (se)[1]
Have you used health care services in the United States in last 2 years?	◆	36.4 (2.3)	11.2 (2.0)	31.0 (6.3)	39.0 (6.0)	47.7 (3.5)
• The last time you used health care services was it:						
Related to your farm work job?	◆	4.2 (0.8)	0.7 (0.3)	5.6 (1.9)	6.9 (1.6)	4.5 (1.2)
Not related to any job?		30.9 (2.2)	10.3 (1.9)	24.4 (5.8)	31.5 (6.5)	41.3 (3.4)
Did not use health care services in last 2 years		63.6 (2.3)	88.8 (2.0)	69.1 (6.3)	61.1 (6.0)	52.3 (3.5)
For those who used health care services in the last 2 years, how did you pay?						
• For those with a problem related to the farm work job?						
Paid self		23.1 (6.0)	31.6 (19.5)	31.2 (18.8)	14.4 (7.0)	26.7 (8.9)
Employer provided health plan or Workers Compensation		59.3 (3.7)	60.6 (19.8)	45.4 (20.4)	58.2 (13.1)	63.8 (10.8)
Other		17.0 (5.1)	7.8 (7.9)	23.4 (12.0)	27.4 (12.2)	9.6 (3.9)
• For those with a problem NOT related to the farm work job?						
Paid self		59.8 (7.8)	67.9 (11.6)	56.4 (10.0)	55.3 (5.7)	60.0 (4.3)
Employer provided health plan or Workers Compensation		9.2 (2.4)	2.6 (2.1)	4.7 (2.5)	3.8 (2.2)	12.2 (3.2)
Other		31.5 (3.0)	29.6 (11.4)	38.8 (11.2)	40.9 (6.2)	27.9 (2.8)
• Is it easy or difficult to get the health care services you need in the United States?	◆					
Difficult		50.5 (2.9)	54.2 (5.9)	57.1 (4.6)	51.1 (3.4)	47.9 (3.6)
Easy		36.8 (3.3)	14.4 (4.4)	37.0 (4.6)	41.0 (4.2)	44.8 (3.6)
Don't know		12.7 (2.1)	31.4 (5.9)	5.9 (1.7)	7.9 (1.7)	7.4 (1.8)
• When was the last time you saw a dentist (In the United States or elsewhere)?						
Never		40.9 (2.7)	61.0 (4.2)	49.9 (4.6)	40.1 (5.0)	30.8 (3.3)

◆ SD=Statistically Different. Rows or groups of rows with an "◆" indicate that differences in prevalence between two or more levels of the stratification variable exist at the p<0.05 level. In cases without an "◆", differences were not statistically significant.

[1] (se) – Standard Error.

Note. Due to rounding, some column totals may not add up to exactly 100 percent.

Chapter 4: Results

Table 27. National Agricultural Workers Survey access to and quality of health care, by number of farmworkers employed on farm, October 1998—September 1999

Health care	SD	Total %(se)[1]	Number of farmworkers employed on farm			
			1–10 %(se)[1]	11–50 %(se)[1]	51–150 %(se)[1]	>150 %(se)[1]
Have you used health care services in the United States in last 2 years?		36.4 (2.3)	46.4 (5.2)	36.6 (3.5)	30.6 (2.7)	39.9 (5.0)
• The last time you used health care services was it:						
Related to your farm work job?		4.2 (0.8)	5.4 (1.0)	6.1 (1.3)	2.6 (0.7)	2.9 (1.2)
Not related to any job?		30.9 (2.2)	39.0 (5.4)	29.0 (3.1)	26.7 (2.2)	36.5 (5.2)
Did not use health care services in last 2 years		63.6 (2.3)	53.6 (5.2)	63.4 (3.5)	69.4 (2.7)	60.1 (5.0)
For those who used health care services in the last 2 years, how did you pay?						
• For those with a problem related to the farm work job?						
Paid self		23.1 (6.0)	16.2 (6.2)	25.5 (8.5)	26.9 (12.9)	14.6 (11.4)
Employer provided health plan or Workers Compensation		59.9 (3.7)	59.3 (14.1)	52.4 (10.7)	67.4 (13.7)	76.0 (15.5)
Other		17.0 (5.1)	24.6 (11.1)	22.1 (7.9)	5.7 (3.4)	9.4 (9.7)
• For those with a problem NOT related to the farm work job?						
Paid self		59.3 (7.8)	59.6 (6.6)	64.4 (6.3)	55.2 (5.5)	56.1 (4.9)
Employer provided health plan or Workers Compensation		9.2 (2.4)	4.7 (1.7)	6.1 (1.7)	11.5 (4.1)	12.2 (5.8)
Other		31.5 (3.0)	35.8 (7.3)	29.6 (5.4)	33.4 (4.3)	31.8 (4.9)
• Is it easy or difficult to get the health care services you need in the United States?	♦					
Difficult		50.5 (2.9)	43.4 (5.4)	43.4 (3.6)	57.1 (5.5)	55.6 (5.6)
Easy		36.8 (3.3)	49.6 (4.9)	46.3 (3.5)	28.0 (4.6)	29.4 (5.4)
Don't know		12.7 (2.1)	7.0 (2.6)	10.3 (1.5)	14.9 (3.5)	15.0 (3.3)
• When was the last time you saw a dentist (In the US or elsewhere)?						
Never		40.9 (2.7)	32.1 (4.8)	43.1 (4.2)	42.2 (4.5)	37.8 (4.1)

♦ SD=Statistically Different. Rows or groups of rows with an "♦" indicate that differences in prevalence between two or more levels of the stratification variable exist at the p<0.05 level. In cases without an "♦", differences were not statistically significant.

[1] (se) – Standard Error.

Note. Due to rounding, some column totals may not add up to exactly 100 percent.

Chapter 4: Results

Table 28. National Agricultural Workers Survey access to and quality of health care, by crop category, October 1998—September 1999

Health care	SD	Total % (se)[1]	Field crops % (se)[1]	Fruit and nuts % (se)[1]	Vege- % (se)[1]	Horti- % (se)[1]	Misc/ % (se)[1]
Have you used health care services in the United States in last 2 years?		36.4 (2.3)	31.1 (4.2)	32.9 (1.8)	34.2 (3.8)	46.3 (6.7)	58.2 (9.3)
• The last time you used health care services was it:							
Related to your farm work job?....................	♦	4.2 (0.8)	3.8 (1.3)	3.7 (1.4)	4.1 (0.7)	5.0 (1.5)	7.9 (4.7)
Not related to any job?.............................		30.9 (2.2)	23.5 (3.4)	28.1 (1.8)	29.9 (3.8)	40.1 (7.5)	50.1 (12.2)
Did not use health care services in last 2 years..		63.6 (2.3)	68.9 (4.2)	67.1 (1.8)	65.8 (3.8)	53.7 (6.7)	41.9 (9.3)
For those who used health care services in the last 2 years how did you pay?							
• For those with a problem related to the farm work job?							
Paid self..		23.1 (6.0)	26.1 (12.3)	7.8 (5.8)	48.9 (12.2)	16.4 (13.6)	21.9 (12.0)
Employer provided health plan or Workers Compensation....................		59.3 (3.7)	56.9 (15.7)	77.8 (10.2)	22.2 (8.4)	68.5 (18.7)	77.7 (12.1)
Other		17.0 (5.1)	17.0 (9.9)	14.4 (8.1)	29.0 (9.8)	15.2 (11.7)	0.4 (0.5)
• For those with a problem NOT related to the farm work job?							
Paid self..		59.3 (7.8)	61.5 (7.6)	63.9 (5.7)	44.8 (9.3)	55.7 (4.4)	87.1 (6.9)
Employer provided health plan or Workers Compensation....................		9.2 (2.4)	6.8 (3.6)	9.6 (4.3)	11.1 (7.7)	11.1 (4.7)	1.1 (1.1)
Other		31.5 (3.0)	31.7 (6.9)	26.6 (3.0)	44.1 (8.0)	33.3 (5.7)	11.8 (6.5)
• Is it easy or difficult to get the health care services you need in the United States?							
Difficult..		50.5 (2.9)	45.5 (5.3)	55.0 (4.2)	53.6 (6.3)	44.8 (5.3)	34.7 (9.7)
Easy		36.8 (3.3)	41.6 (4.9)	30.2 (4.0)	36.5 (8.0)	43.2 (4.7)	53.2 (11.5)
Don't know...		12.7 (2.1)	12.9 (3.2)	14.8 (4.0)	9.9 (2.3)	12.0 (3.0)	12.2 (4.5)
• When was the last time you saw a dentist (In the United States or elsewhere)?							
Never		40.9 (2.7)	50.8 (5.2)	38.4 (4.0)	48.8 (4.8)	29.0 (4.7)	29.8 (7.6)

♦ SD=Statistically Different. Rows or groups of rows with an "♦" indicate that differences in prevalence between two or more levels of the stratification variable exist at the p<0.05 level. In cases without an "♦", differences were not statistically significant.

[1] (se) – Standard Error.

Note. Due to rounding, some column totals may not add up to exactly 100 percent.

Chapter 4: Results

Section Seven: Estimated Prevalence of Physician-Diagnosed Health Conditions

Table 29 shows the prevalence of physician diagnosed health conditions. Approximately 12% of farmworkers reported one or more of these conditions. The highest reported physician diagnosed health condition was high blood pressure (4%). Because of low prevalence, these conditions were not stratified by the farmworker and farm characteristics.

Table 29. National Agricultural Workers Survey estimated prevalence of physician-diagnosed health conditions, October 1998—September 1999

Health condition	Total %	(se)[1]
Have you ever been told by a doctor or nurse that you have any health condition?		
• Type of health condition*		
Any	12.1	(1.3)
Asthma	1.8	(0.5)
Cancer	0.1	(0.5)
Diabetes	2.0	(0.4)
Hepatitis	0.2	(0.1)
High blood pressure	3.5	(0.5)
Heart disease	0.8	(0.2)
Thyroid disease	0.6	(0.5)
Tuberculosis	0.8	(0.5)
Urinary tract infection	1.9	(0.9)
Other	2.3	(0.6)

*Respondents may have reported more than one health condition.
[1] (se) – Standard Error.

References

Alderete E, Vega WA, Kolody, Bohadan, Aguilar-Gaxiola S [1999]. Depressive symptomatology: prevalance and psychosocial risk factors among Mexican migrant farmworkers in California. J Community Psychol 27(4):457–471.

Arcury TA, Quandt SA [2001]. Farmworker pesticide exposure and community-based participatory research: rationale and practical applications. Environ Health Perspect 109(Suppl 3):429–434.

Austin C, Arcury TA, Quandt SA, Preisser JS, Saavedra RM, Cabrera LF [2001]. Training farmworkers about pesticide safety: issues of control. J Health Care Poor Underserved 12(2):236–249.

Babubhai VS, Barnwell BG, Bieler GS [1997]. SUDAAN user's manual, Release 7.5. Research Triangle Park, NC: Research Triangle Institute.

Ballard T, Ehlers J, Freund E, Auslander M, Brandt V, Halperin W [1995]. Green tobacco sickness: occupational nicotine poisoning in tobacco workers. Arch Environ Health 50(5):384–389.

Beaumont JJ, Goldsmith DF, Morrin LA, Schenker MB [1995]. Mortality in agricultural workers after compensation claims for respiratory disease, pesticide illness and injury. J Occup Environ Med 37(2):160–169.

BLS [2002]. Occupational safety and health summary data. Washington, DC: U.S. Department of Labor, Bureau of Labor Statistics. Available at: www.bls.gov/iif/oshsum1.htm

BLS [2005]. Census of Fatal Occupational Injuries Charts 1992-2005. Washington, DC: U.S. Department of Labor, Bureau of Labor Statistics. Available at: http://www.bls.gov/iif/oshwc/cfoi/cfch0004.pdf

Brackbill RM, Cameron LL, Behrens V [1994]. Prevalence of chronic diseases and impairments among US Farmers, 1986–1990. Am J Epidemiol 139:1055–1065.

Bradman MA, Harnly ME, Draper W, Seidel S, Teran S, Wakeham D, Neutra R [1997]. Pesticide exposures to children from California's Central Valley: results of a pilot study. J Expo Anal Environ Epidemiol 7(2):217–234.

Ciesielski SD, Seed JR, Esposito DH, Hunter N [1991]. The epidemiology of tuberculosis among North Carolina migrant farm workers. JAMA 265(13):1715–1719.

Ciesielski SD, Seed JR, Ortiz JC, Metts J [1992]. Intestinal parasites among North Carolina migrant farmworkers. Am J Public Health 82(9):1258–1262.

References

Di Natale M [2002]. Personal communication from M Di Natale, BLS Economist, Current Population Survey, Bureau of Labor Statistics, Washington, DC, September 13.

DOL [2000]. U.S. Department of Labor report to Congress: the agricultural labor market–status and recommendations. Available at:
http://migration.ucdavis.edu/rmn/word-etc/dec_2000_labor.htm

DOL [2001]. Proposed information collection request submitted for public comment and recommendations: The National Agricultural Workers Survey. Washington, DC: U.S. Department of Labor. Available at: www.dol.gov/asp/regs/fedreg/notices/2001019311.htm

DOL [2002]. Comparison of State unemployment laws. Washington, DC: U.S. Department of Labor, Employment and Training Administration. Available at:
http://workforcesecurity.doleta.gov/unemploy/comparison2002.asp

DOL. The National Agricultural Workers Survey (NAWS). Washington, DC: U.S. Department of Labor. Available at: www.dol.gov/asp/programs/agworker/naws.htm.

EPA [1993]. The worker protection standard for agricultural pesticides: how to comply. Available at: www.epa.gov/cgi-bin/claritgw?op-Display&document=clserv:OPPTS:0079;&rank=4&template=epa

EPA [1997]. The worker protection standard and recent amendments. Washington, DC: U.S. Environmental Protection Agency. Available at:
www.epa.gov/oppfead1/safety/workers/amendmnt.htm.

EPA [1999]. Who and what are covered? Washington, DC: U.S. Environmental Protection Agency. Available at: www.epa.gov/pesticides/safety/workers/awsscope.htm.

Fenske RA [1997]. Pesticide exposure assessment of workers and their families. Occup Med: State of the Art Reviews 12(2): 221–237.

Franzini L, Ribble JC [2001]. Understanding the Hispanic paradox. Ethn Dis 11(3):496–518.

JTPA. Gloucester County Employment and Training Agency, Job Training Partnership Act. Available at: www.co.gloucester.nj.us/jtpa/.

Hernberg S [1992]. Introduction to occupational epidemiology. Boca Raton, FL: Lewis Publishers.

Hovey J, Magaña C [2002]. Psychosocial predictors of anxiety among immigrant Mexican migrant farmworkers: implications for prevention and treatment. Cult Divers Ethnic Minor Psychol 8(3):274–289.

References

Hovey J, Magana C [2003]. Cognitive, affective, and physiological expressions of anxiety symptomatology among Mexican migrant farmworkers: predictors and generational differences. Community Ment Health J 38(3):223–237.

Last JM [1995]. A dictionary of epidemiology. New York: Oxford University Press.

Leigh PJ, McCurdy, SA, Schenker MB [2001]. Costs of occupational injuries in agriculture. Public Health Reports 116:235–248.

McBride DI, Firth HM, Herbison GP [2003]. Noise exposure and hearing loss in agriculture: a survey of farmers and farm workers in the Southland region of New Zealand. J Occup Environ Med 45(12):1281–1288.

McCurdy SA, Carroll DJ [2000]. Agricultural injury. 38(4):463–480.

Mehta K, Gabbard SM., Barrat V, Lewis M, Carroll D, Mines R [2000]. Findings from the National Agricultural Workers Survey (NAWS) 1997–1998: a demographic and employment profile of United States farmworkers (Research Report No. 8). Washington, DC: U.S. Department of Labor. Available at: http://www.doleta.gov/agworker/naws.cfm

Meister JS [1991]. The health of migrant farm workers. Occup Med: State of the Art Reviews 6(3):503–518.

Merchant JA, Kross BC, et al. [1989]. Agriculture at risk: a report to the Nation (Agricultural occupational and environmental health: policy strategies for the future). Iowa City and Des Moins, IA: National Coalition for Agricultural Safety and Health, Institute of Agricultural Medicine and Occupational Health.

Mines R, Alarcon R, Mehta K [1999]. Family separation: the changing pattern of Mexican migration to U.S. agriculture. Denver, CO: National Association of Community Health Centers National Migrant Health Meeting, April 23.

Moses M, Johnson ES, Anger WK, Burse VW, Horstman SW, Jackson RJ, Lewis RG, Maddy KT, McConnell R, Meggs WJ, et al. [1993]. Environmental equity and pesticide exposure. Environ Ind Health 9:913–959.

NCFH [1985–2002]. Overview of America's farmworkers: occupational safety and health. Buda, TX: National Center for Farmworker Health. Available at: www.ncfh.org/aaf_03.php.

NSC [2000]. Report on injuries in America. Itasca, IL: National Safety Council. Available at: www.nsc.org/lrs/statinfo/99report.htm.

References

OSHA [1992]. Field Sanitation Standard. Washington, DC: U.S. Department of Labor, Occupational Safety and Health Administration. Available at: www.osha.gov/pls/oshaweb/owadisp.show_document?p_table=FACT_SHEETS&p_id=137

O'Connor T [2003]. Reaching Spanish-speaking workers and employers with occupational safety and health information. In: Safety is seguridad: a workshop summary. Washington, DC: National Academies Press.

Rosenman KD, Kalush A, Reilly MJ, Gardiner JC, Reeves M, Luo Zhewui [2006]. How much work-related injury and illness is missed by the current national surveillance system. JOEM 48(4): 357-365.

Schenker MB [1996]. Preventive medicine and health promotion are overdue in the agricultural workplace. J Public Health Policy 17(3):275–305.

Sherman J, Villarejo D, Garcia A, McCurdy S, Mobed K, Runsten D, Saiki K, Samuels S, Schenker M [1997]. Finding invisible farm workers: The Parlier Survey. Davis, CA: California Institute for Rural Studies. Summary available at: www.cirsinc.org/pub/parlier.html

Shuval JT [1993]. Migration and stress. In: Goldberger L, Breznitz S, eds. Handbook of stress: theoretical and clinical aspects. New York: The Free Press, pp. 641–657.

USDA [1988]. Temporary foreign worker program—summary. Washington, DC: U.S. Department of Agriculture. Available at: www.usda.gov/agency.oce/oce/labor-affairs/h2asumm.htm.

Villarejo D [2003]. The health of U.S. hired farm workers. Ann Rev Public Health 24:175–193.

Villarejo D, Lighthall D, Williams III D, Souter A, Mines R, Bade B, Samuels S, McCurdy SA [2001]. Suffering in Silence: A report on the health of California's Agricultural Workers. California Institute for Rural Studies, Davis California.

Villarejo D, Baron S [1999]. The occupational health status of hired farm workers. Occup Med: State of the Art Reviews. 14(3):613–635.

Weathers A, Minkovitz C, O'Campo P, Diener-West M [2004]. Access to care for children of migratory agricultural workers: factors associated with unmet need for medical care. Pediatrics. 113(4):276–282.

Wilk V, Holden R [1998]. New directions in the surveillance of hired farm worker health and occupational safety: a report of the work group convened by NIOSH (May 5, 1995) to identify priorities for hired farm worker occupational health surveillance and research. Cincinnati, OH: U.S. Department of Health and Human Services, Centers for Disease Control and Prevention, National Institute for Occupational Safety and Health. Available at: www.cdc.gov/niosh/hfw-index.html.

Appendix A

Survey Instrument

Introduction to Appendix A

The purpose of this appendix is to give users of this publication easy access to the Survey Instrument.

How to Use this appendix

Appendix A contains the Survey Instrument in English. The majority of the respondents were interviewed in Spanish. The Spanish language Survey Instrument can be viewed at the publicatons section of the NIOSH web sitewww.cdc.gov/niosh/docs/2009-119/.

Appendix A

Rev. 6/15/99 English Ver. 2
Cycle 34, Summer 1999
OMB 1225-0044

COUNTY	FARMWORKER ID
	3 4

[FOR OFFICE USE ONLY]

NATIONAL AGRICULTURAL WORKERS SURVEY 1999

CS2 DATE: ☐☐ / ☐☐ / ☐☐

[FOR OFFICE USE ONLY]
CROP CODE
TASK CODE

CS5 CROP:

CS6 TASK:

WORKER IS ACTUALLY EMPLOYED BY:

○ GROWER ○ CONTRACTOR ○ NURSERY ○ PACKING HOUSE ○ OTHER

GN: _____ **ID:** ☐☐☐☐☐☐☐☐

From List? ○ Yes ○ No

FARMWORKER'S NAME:

LOCAL STREET ADDRESS:

MAILING ADDRESS:

PHONE NUMBER: **HOME:** _____ **MESSAGE:** _____

NAME OF INTERVIEWER: _____ **CS9 INTERVIEWER ID:** ☐☐☐

CP5 TIME BEGAN: ☐☐:☐☐ ☐ AM ☐ PM   **CP6 TIME ENDED:** ☐☐:☐☐ ☐ AM ☐ PM

Public reporting burden for the collection of information is estimated to average 1 hour (or 60 minutes) per response, including the time for reviewing instructions, searching existing data sources, gathering and maintaining the data needed, and completing and reviewing the collection of information. Persons are not required to respond to the collection of information unless it displays a currently valid OMB control number. Send comments regarding this burden estimate or any other aspect of this collection of information, including suggestions for reducing this burden, to the Office of Information Management, Department of Labor, Room N-1301, 200 Constitution Avenue, N.W., Washington, D.C. 20210; and to the Office of Information and Regulatory Affairs, Office of Management and Budget, Washington, D.C. 20503.

Appendix A

HOUSEHOLD GRID

County _____ Farmworker ID _____

REFER TO QUESTIONS IN SECTION A:

A1	A2	A3	A4	A5	A6	A7	A8	A9	A10	A11	A11a	A12	A13
NAME (FARMWORKER) DATE OF MARRIAGE/UNION (MM/YY)	RELATION	SEX	DOES HE/SHE LIVE WITH YOU NOW? IF NOT WHERE?	MARITAL STATUS	BIRTH DATE MM/YY	PLACE OF BIRTH	YEAR ENTERED US	HIGHEST GRADE	COUNTRY SCHOOL	ANY US SCHOOL LAST 12 MONTHS?	ANY SCHOOL NOW?	WORK	ANY US FARM WORK LAST 12 MONTHS?
		M F		S M O									
		M F	Y N	S M O						Y N N/A	Y N N/A	FW NF NW	Y N N/A
		M F	Y N	S M O						Y N N/A	Y N N/A	FW NF NW	Y N N/A
		M F	Y N	S M O						Y N N/A	Y N N/A	FW NF NW	Y N N/A
		M F	Y N	S M O						Y N N/A	Y N N/A	FW NF NW	Y N N/A
		M F	Y N	S M O						Y N N/A	Y N N/A	FW NF NW	Y N N/A
		M F	Y N	S M O						Y N N/A	Y N N/A	FW NF NW	Y N N/A
		M F	Y N	S M O						Y N N/A	Y N N/A	FW NF NW	Y N N/A

CODES FOR A2:
1 = Spouse/common law spouse
2 = Own child
3 = Sibling
4 = Parent
5 = Grandchild
6 = Other relative (cousins, uncles, etc.)
7 = Other

(COUNTRY CODES) FOR A7 AND A10:
1 = USA
2 = Puerto Rico
3 = Mexico
4 = Central America
5 = South America
6 = Caribbean
7 = Southeast Asia (Indonesia, Cambodia, Vietnam, Laos, Thailand)
8 = Pacific Islands (The Philippines, Guam, Fiji, etc.)
9 = Asia (China, Japan, Korea, etc.)
97 = Other
99 = Not answered

Appendix A

HOUSEHOLD GRID

REFER TO QUESTIONS IN SECTION A:

County _____ Farmworker ID _____

A1	A2	A3	A4	A5	A6	A7	A8	A9	A10	A11	A11a	A12	A13
NAME	RELATION	SEX	DOES HE/SHE LIVE WITH YOU NOW? IF NOT WHERE?	MARITAL STATUS	BIRTH DATE MM/YY	PLACE OF BIRTH	YEAR ENTERED U.S.	HIGHEST GRADE	COUNTRY SCHOOL	ANY U.S. SCHOOL LAST 12 MONTHS?	ANY SCHOOL NOW?	WORK	ANY U.S. FARM WORK LAST 12 MONTHS?
		M F	Y N	S M O						Y N N/A	Y N N/A	FW NF NW	Y N N/A
		M F	Y N	S M O						Y N N/A	Y N N/A	FW NF NW	Y N N/A
		M F	Y N	S M O						Y N N/A	Y N N/A	FW NF NW	Y N N/A
		M F	Y N	S M O						Y N N/A	Y N N/A	FW NF NW	Y N N/A
		M F	Y N	S M O						Y N N/A	Y N N/A	FW NF NW	Y N N/A
		M F	Y N	S M O						Y N N/A	Y N N/A	FW NF NW	Y N N/A
		M F	Y N	S M O						Y N N/A	Y N N/A	FW NF NW	Y N N/A

CODES FOR A2:
1 = Spouse/common law spouse
2 = Own child 3 = Sibling
4 = Parent
5 = Grandchild
6 = Other relative (cousins, uncles, etc.)
7 = Other:

(COUNTRY CODES) FOR A7 AND A10:
1 = U.S.A.
2 = Puerto Rico
3 = Mexico
4 = Central America
5 = South America
6 = Caribbean
7 = Southeast Asia (Indonesia, Cambodia, Vietnam, Laos, Thailand)
8 = Pacific Islands (The Philippines, Guam, etc.)
9 = Asia (China, Japan, Korea, etc.)
97 = Other:

99 = Not answered

Please answer the following questions regarding other individuals who live with you, are not your relatives and were not mentioned earlier.

A15   Other than those you have already mentioned, how many people live with you now?

[Total] [ ]

[A15] Out of those How many are............? [WRITE TOTAL BELOW]		[A16] # doing FW	[A17] # doing NF	[A18] # doing NW
a. Adults: 18 years or older				
b. Children: 17 years old or younger				
c. Don't Know Age				

[ONLY FOR THOSE WHO WORK IN BORDER CITIES (WITH MEXICO)]

A19.   Do you commute across the border for your FW days?

☐ 0 No   ☐ 1 Yes

Appendix A

[THE FOLLOWING QUESTIONS ARE SOLELY FOR THE INTERVIEWEE.]

B1 Which of the following describes you? [READ CHOICES. MARK ONE RESPONSE.]

- ○ 1 Mexican-American
- ○ 2 Mexican
- ○ 3 Chicano
- ○ 5 Puerto Rican
- ○ 4 Other Hispanic: _____
- ○ 7 Not Hispanic or Latino
- ○ 9 Not answered

B2 Which of the following do you consider yourself? [READ CHOICES. MARK ONE RESPONSE.]

- ○ 1 White
- ○ 2 Black or African American
- ○ 4 American Indian, Alaskan Native, (Indigenous)
- ○ 5 Asian
- ○ 6 Native Hawaiian or Pacific Islander
- ○ 7 Other: _____
- ○ 9 Not answered

B3 Have you attended any of the following special classes or school in the U.S.? [READ CHOICES. MARK ALL THAT APPLY.]

- ○ a. English
- ○ b. Citizenship
- ○ c. Literacy
- ○ d. Job training: Note: _____
- ○ e. GED, (High School Equivalency)
- ○ f. College or University
- ○ g. Adult Basic Education
- ○ h. Even Start
- ○ i. Migrant Education
- ○ j. Other: _____
- ○ None
- ○ Not answered

B5 What is your first or primary language? [DO NOT READ CHOICES. MARK ONE RESPONSE. IF RESPONDENT IS CONFUSED, PROMPT with: "What language do you speak at home?"]

- ○ 1 English
- ○ 2 Spanish
- ○ 3 French
- ○ 4 Creole
- ○ 5 Laotian
- ○ 6 Hmong
- ○ 7 Vietnamese
- ○ 8 Cambodian
- ○ 9 Tagalog/Ilocano
- ○ 10 Mixtec
- ○ 11 Kanjobal
- ○ 97 Other: _____
- ○ 99 Not answered

B6 [IF PRIMARY LANGUAGE IS NOT ENGLISH] How well do you read in your primary language? [IF THE LANGUAGE DOES NOT HAVE A WRITTEN FORM, ASK ABOUT LANGUAGE USED IN SCHOOL.]

- ○ 1 Not at all
- ○ 2 A little
- ○ 3 Somewhat
- ○ 4 Well

B7 How well do you speak English? [READ CHOICES. MARK ONE RESPONSE.]

- ○ 1 Not at all
- ○ 2 A little
- ○ 3 Somewhat
- ○ 4 Well

B8 How well do you read English? [READ CHOICES. MARK ONE RESPONSE.]

- ○ 1 Not at all
- ○ 2 A little
- ○ 3 Somewhat
- ○ 4 Well

B10 In what year did you first do any farm work in the U.S.?

[1][9][ ][ ]

B11 Approximately how many years have you done FARM WORK in the U.S.? [COUNT ANY YEAR IN WHICH 15 DAYS OR MORE WERE WORKED.]

[ ][ ] years

B12 Approximately how many years have you done NON-FARM WORK in the U.S.? [COUNT ANY YEAR IN WHICH 15 DAYS OR MORE WERE WORKED.]

[ ][ ] years

B13 When was the last time your parents did farm work in the U.S.?

○ 0 Never
○ 1 Now/within last year
○ 2 One to five years ago
○ 3 Six to ten years ago
○ 4 Over 11 years ago
○ 7 Don't know

B14 [ASK ALL] What state do you consider to be your permanent residence (i.e. home)? [IF IT IS IN A FOREIGN COUNTRY, ENTER STATE, DEPARTMENT, OR PROVINCE. IF NO PERMANENT HOME, WRITE "NONE".]

[ ][ ][ ][ ][ ][ ][ ][ ][ ][ ]

B15 Before coming to this state [name of state], in what state did you live? (In the U.S.)

[ ][ ][ ][ ][ ][ ][ ][ ][ ][ ]

B16 [IF FOREIGN BORN] When you lived in your country (outside the U.S.), did you work in ... ? [READ CHOICES. MARK ONE RESPONSE.]

○ 1 Agriculture
○ 2 Non-agriculture (NF)
○ 3 Part farm and part non farm
○ 5 Never worked
○ 7 Other: [_____]
○ 8 Not applicable (Only for those born in the U.S.)
○ 9 Not answered

B17 [IF FOREIGN BORN] In what country (outside of the U.S.) did you live before coming to the U.S.?

[ ][ ][ ][ ][ ][ ][ ][ ][ ][ ][ ][ ]

B18 [IF FOREIGN BORN] Before coming to the United States, in what state/department/ province did you live?

[ ][ ][ ][ ][ ][ ][ ][ ][ ][ ][ ][ ]

Appendix A

Work Grid

REFER TO QUESTIONS IN THE FOLLOWING SECTION

[C1-C2 FOR OFFICE USE ONLY] REPORT ONLY FROM FIRST PERIOD COVERING JUNE 1, 1998 TO PRESENT

County _____ Farmworker ID _____

C1-C2	C15	C3	C4	C5	C6	C8	C9		C10	C11	C12	C13	C7	C16
PER. AND SUB PER. NO.	GR CO DK NA	EMPLOYER (FARM WORK OR NON-FARM JOB)	CROP	ACTIVITY/ TASK, WHILE FW NF OR *NW	FW NF NW AB	UNEMPLOYMENT BENEFITS	DATES FOR PERIODS OF FW,NF, NW,AB FROM:	TO:	FW AND NF DAYS PER WEEK	CITY	COUNTY	STATE/COUNTRY	WHY LEFT? **FW AND NF	DID YOUR SPOUSE & KIDS MOVE WITH YOU
	GR CO DK NA				FW NF NW AB	Y N								SPOUSE CHILDREN ALL NO NA
	GR CO DK NA				FW NF NW AB	Y N								SPOUSE CHILDREN ALL NO NA
	GR CO DK NA				FW NF NW AB	Y N								SPOUSE CHILDREN ALL NO NA
	GR CO DK NA				FW NF NW AB	Y N								SPOUSE CHILDREN ALL NO NA
	GR CO DK NA				FW NF NW AB	Y N								SPOUSE CHILDREN ALL NO NA

*C-5 CODES: ONLY FOR ACTIVITY WHILE NOT WORKING. (Write activity for FW and NF)

201 = LOOKING FOR FARMWORK AND NON-FARMWORK
202 = LOOKING FOR FARM WORK
203 = LOOKING FOR NON-FARM WORK
204 = WAITING FOR RECALL NOTICE (AFTER LAYOFF)
205 = WAITING FOR START OF SEASON
206 = FAMILY RESPONSIBILITIES WORK IN HOME
207 = IN SCHOOL
208 = LAID UP DUE TO INJURY
209 = IN-TRANSIT BETWEEN JOBS
210 = VACATION
211 = DID NOT LOOK FOR WORK
212 = OTHER (SPECIFY IN BOX)

**CODES FOR C-7

1 = LAID OFF/END OF SEASON
2 = FIRED
3 = FAMILY RESPONSIBILITIES
4 = SCHOOL
5 = MOVED
6 = HEALTH REASON
7 = VACATION
8 = RETIRED
9 = OTHER (SPECIFY)
10 = QUIT
11 = CHANGE JOBS

A8

Appendix A

Pages 9 through 11 same as page 8

D1 [SHOW CALENDAR]

In the year before last, [THE YEAR BEFORE THE ONE COVERED IN THE WORK GRID] how many months did you do (FW) in the U.S.? [1 DAY OR MORE PER MONTH EQUALS 1 MONTH.]

☐☐ Months

D2 [IF NON-FARM JOB LISTED ON WORK GRID:] For your most recent non-farm (NF) employer, how many hours per week did you work on average?

☐☐☐ Hours

D3 [IF NON-FARM JOB LISTED:] For your most recent non-farm employer, how much were you paid per week on average?

$ ☐,☐☐☐.☐☐

"CURRENT FARM"

Now I am going to ask you some questions about the crop/task you are CURRENTLY performing for the EMPLOYER through whom we contacted you [Grower list employer]. [IF RESPONDENT INDICATES MORE THAN ONE CROP/TASK, ASK FOR THE ONE HE/SHE DOES THE MOST.]

D4 How many hours did you work last week at your current farm job?

☐☐☐

D9 You already told me that the crop you are currently working is:

☐☐☐☐☐☐☐☐

D10 You already told me that the task you are currently doing is:

☐☐☐☐☐☐☐☐

[D5 TO D8: IF HE/SHE HAS NOT RECEIVED PAYMENT YET FOR CURRENT CROP, ASK FOR ESTIMATES.] Can you tell me how you were paid and the amount your employer paid you on your last pay day?

D5 After taxes:

$ ☐,☐☐☐.☐☐

D6 Before taxes:

$ ☐,☐☐☐.☐☐

D61 Are you paid by: [READ CHOICES. MARK ONE RESPONSE.]

☐ 1 Payroll Check? ☐ 4 Other check?
☐ 2 Personal check? ☐ 5 Cash?
☐ 3 Cash and check? ☐ 6 Other: ☐

D62 Did you get a receipt?

☐ 1 Yes ☐ 0 No

D7 For what time period was that payment?

○ 1 One day ○ 2 One week ○ 3 Two weeks
○ 4 One month ○ 7 Other: ☐

D8 How many hours did you work during that period (in D7)?

☐☐☐ Hours

D11 Are you paid:

○ 1 By the hour?
○ 2 By the piece? [SKIP TO D13]
○ 3 Combination hourly wage and piece rate?
   [ASK D12 THRU D18]
○ 4 Salary or other? [SKIP TO D19]

12

Appendix A

D12 [IF PAID BY THE HOUR:] How much per hour (to nearest cent)?

$ ☐☐ . ☐☐

D13 [IF PAID BY THE PIECE:] Are you paid as an individual or by the crew? [IF THE ANSWER IS "CREW", ASK QUESTIONS D14 to D18 CONSISTENTLY IN REFERENCE TO THE CREW.]

○ 1 Individual (SKIP TO D15)
○ 2 Crew

D14 [IF CREW PIECE RATE:] How many people are in your crew? [ONE IS NOT A POSSIBLE ANSWER.]

☐☐☐

D15 [IF BY PIECE:] How do they pay you/your crew? [i.e., UNIT OF MEASURES SUCH AS BOX, BIN, BUCKET, ETC.]

☐☐☐☐☐☐☐☐☐☐

D16 [IF BY PIECE:] How many of these (boxes, bins, buckets, etc.) do you/your crew do in an average day?

☐☐☐

D17 [IF BY PIECE:] How many hours per day do you/your crew work on average at this task?

☐☐ Hours

D18 [IF BY PIECE:] How much do they pay you/your crew on average for each box bin, bucket, etc. (in D15)?

$ ☐ , ☐☐☐ . ☐☐

D19 [IF PAID BY SALARY, OR OTHER:] Explain fully how and how much you are paid (salary or other). Explain thoroughly the method and amount of payment.

D20 Aside from your wages, do you receive any other money bonus from your employer?

○ 0 No [SKIP TO D22]
○ 1 Yes
○ 7 Don't know [SKIP TO D22]
○ 9 Not answered [SKIP TO D22]

D21 [IF PAID A BONUS:] How and when do you receive the bonus? [READ CHOICES. MARK ALL THAT APPLY.]

○ a. Holiday bonus
○ b. Incentive bonus (rewards)
○ c. Dependent on grower profit
○ d. End of season bonus
○ e. Money for transportation
○ f. Other ☐

D63 How much were you given (TOTAL)?

$ ☐ , ☐☐☐ . ☐☐

13

Appendix A

D22 If you are injured AT WORK or get sick as a result of your work, does your employer provide health insurance or pay for your health care?

- O 0 No
- O 1 Yes
- O 7 Don't Know
- O 9 Not answered

D23 If you are injured AT WORK or get sick as a result of your work, do you get any payment while you are recuperating (i.e., workers' compensation)?

- O 0 No
- O 1 Yes
- O 7 Don't Know
- O 9 Not answered

D24 If you are injured or get sick OFF THE JOB (e.g., at home), does your employer provide health insurance or pay for your health care?

- O 0 No
- O 1 Yes
- O 7 Don't Know
- O 9 Not answered

D26 Are you covered by unemployment insurance if you lose this job?

- O 0 No
- O 1 Yes
- O 7 Don't Know
- O 9 Not answered

D27 How many years have you worked for this employer? [ONE DAY/PER YEAR=ONE YEAR]

☐☐

D29 [IF WORKED ON A SEASONAL BASIS AND LAID OFF WHEN THE SEASON ENDED] Does this employer keep in contact with you about future employment? [READ CHOICES. MARK ALL THAT APPLY.]

- O a. Yes, before leaving at the end of the season
- O b. Yes, by letter (written message)
- O c. Yes, by phone/in person
- O d. Yes, by someone else
- O e. No, I contact employer
- O f. Other: ☐
- O Don't know

D30 How did you get this job (the first time)? [DO NOT READ CHOICES. MARK ONE RESPONSE.]

- O 1 I applied for the job on my own
- O 4 I was recruited by a grower or his foreman
- O 5 I was recruited by farm labor contractor or his foreman
- O 6 I was refered by the employment services
- O 7 I was refered by the welfare office
- O 8 I was refered by relative/friend/workmate
- O 9 I was referred by labor union
- O 10 Day Laborer/Picked up at Shape Up
- O 97 Other: ☐
- O 99 Not answered

Appendix A

D33a While you are working for this grower/ contractor, what type of arrangement do you have for your living quarters? [DO NOT READ CHOICES. MARK ONLY ONE RESPONSE.]

- ○ 1 I receive free housing from my employer. I PAY NO RENT (pay only a nominal fee for utilities not counted as rent)
- ○ 2 MY FAMILY AND I receive free housing from my employer. I PAY NO RENT (I pay only a nominal fee for utilities not counted as rent)
- ○ 3 I pay for housing provided by my employer. I pay directly or through wage deduction.
- ○ 4 I receive free housing provided by the government, a charity, or other non-work related institution. [I PAY NO RENT]. I pay only a nominal fee for utilities.)
- ○ 5 I pay for housing provided by the government, a charity, or other non-work related institution.
- ○ 6 I (or a family member) own the house.
- ○ 7 I rent from non-employer
- ○ 97 Other: _____

D34 In what type of living quarters do you live now (at this location)? [READ CHOICES. MARK ONLY ONE RESPONSE.]

- ○ 1 House
- ○ 2 Flat or apartment
- ○ 3 Room in hotel, motel, etc.
- ○ 4 Room /bed in rooming/dormitory/boarding house
- ○ 5 Mobile home or trailer (fixed/trailer parks)
- ○ 6 Vehicle (recreational vehicle - RV/camper)
- ○ 7 Homeless (lives outdoors, in a car, tent, lean-to, under bridge or elsewhere with no fixed shelter) [SKIP TOD36a]
- ○ 97 Other: _____

D35 Where are your living quarters located? READ CHOICES. MARK ONLY ONE RESPONSE.]

- ○ 1 Off farm (property not owned/administered by present employer
- ○ 2 Off farm (property owned/administered by present employer
- ○ 3 On farm of the grower I currently work for
- ○ 7 Other: _____

D50 At this location how much do YOU pay for housing (including housing for your family, if they live with you)?

- ○ 1 $ per week ☐,☐☐☐.☐☐
- $ per month ☐,☐☐☐.☐☐
- $ per day ☐,☐☐☐.☐☐
- ○ 2 Don't know, taken out of my paycheck
- ○ 3 Don't know/don't remember, but NOT taken out of my paycheck
- ○ 8 Free housing
- ○ 7 Other: _____

D51 How much is the rent for the entire house/apt/trailer?

- ○ 1 $ per week ☐,☐☐☐.☐☐
- $ per month ☐,☐☐☐.☐☐
- $ per day ☐,☐☐☐.☐☐
- ○ 2 Don't know, taken out of my paycheck
- ○ 3 Don't know/don't remember, but NOT taken out of my paycheck
- ○ 8 Free housing
- ○ 7 Other: _____

Appendix A

D53 In your current living quarters, how many rooms are used for sleeping?

☐☐

D52 How many people total sleep in these rooms?

☐☐☐

D36a [FOR PARENTS OF CHILDREN AGE 12 OR UNDER] During the past 12 MONTHS, where have your children, 12 and under, been while you work in U.S. farm work? [CHECK ALL THAT APPLY.]

○ 1 They've stayed home alone, at least sometimes

○ 13 With my spouse, other family

○ 14 With a neighbor/babysitter, Migrant Head Start, With Head Start, Migrant Education, daycare center etc.

○ 11 With me in the fields

○ 12 Other: _____

D36c [FOR PARENTS OF CHILDREN AGE 12 OR YOUNGER] In the last 12 MONTHS, have any of your children under 12 years old, accompanied you in the fields as you work in the U.S.? [INCLUDE "SOMETIMES" AND MARK ALL THAT APPLY.]

○ 0 No, never

○ 1 Yes, under age 5

○ 2 Yes, between ages 5 and 12

○ 8 Not applicable

D37a How far is your current job from your current residence?

○ 1 I'm located at the job

○ 2 Within 9 miles

○ 3 10-24 miles

○ 4 25-49 miles

○ 5 50-74 miles

○ 6 75 miles or more

D37 At your current job, how do you usually get to work? [READ CHOICES. MARK ONE RESPONSE.]

○ 1 Drive car [SKIP TO D39a]

○ 2 Walk [SKIP TO D39a]

○ 4 Ride with others

○ 5 Public transportation (bus, train) [SKIP TO D39a]

○ 6 Labor bus/truck/van

○ 7 Other: _____

○ 8 "Raitero"

D38a Do you have to use the (transport in D37)? (IS IT OBLIGATORY)?

○ 0 No

○ 1 Yes

○ 7 Don't Know

○ 9 Not answered

16

A14

D38   Do you pay a fee to (responsible in D37), "raiteros" for rides to work?

- O 0 No [SKIP TO D39a]
- O 1 Yes
- O 2 Yes, just for gas
- O 7 Don't know [SKIP TO D39a]
- O 9 Not answered [SKIP TO D39a]

D38b   [ASK ONLY IF THE ANSWER IS "YES" ON D38:] How much do you pay per day or per week?

Per day $ ☐☐☐ . ☐☐

Per week $ ☐☐☐ . ☐☐

D39a   At your current job, who pays for the tools you use at work? [READ CHOICES. MARK ONE RESPONSE.]

- O 1   I don't need any tools [SKIP TO E1]
- O 2   I pay all
- O 3   The grower/contractor [SKIP TO E1]
- O 5   A friend/relative
- O 6   I pay some
- O 10 I pay only for replacement of damaged tools
- O 97 Other: ☐

D39b   How much was paid for equipment at current job, or if you have been at your current job more than 1 year, how much was paid in the last 12 months?

$ ☐ , ☐☐☐ . ☐☐

E1   At any time in the last two years in the U.S.A., were you covered by a union contract while doing farm work?

- O 0 No
- O 1 Yes
- O 7 Don't Know
- O 9 Not answered

E2   How long do you expect to continue doing farm work in the U.S.A.? [READ CHOICES. MARK ONE RESPONSE.]

- O 1 Less than one year
- O 2 One to three years
- O 3 Four to five years
- O 4 Over five years
- O 5 Over five years and as long as I am able
- O 7 Other: ☐
- O 9 Not answered

E3   Do you have any relatives/close friends who work in non-farm work in the U.S.A.?

- O 0 No
- O 1 Yes
- O 7 I don't know
- O 9 Not answered

E4   Could you get a U.S.A. NON-FARM JOB within a month? [READ CHOICES. MARK ONE RESPONSE.]

- O 0 No
- O 1 Yes
- O 7 I don't know
- O 9 Not answered

Appendix A

G1. What was your TOTAL INCOME last year in U.S. dollars (U.S. earnings only)? [READ/SHOW CHOICES. MARK ONE RESPONSE.]

- ○ 1 Under than 500
- ○ 2 500 a 999
- ○ 3 1,000 a 2,499
- ○ 4 2,500 a 4,999
- ○ 5 5,000 a 7,499
- ○ 6 7,500 a 9,999
- ○ 7 10,000 a 12,499
- ○ 8 12,500 a 14,999
- ○ 9 15,000 a 17,499
- ○ 10 17,500 a 19,999
- ○ 11 20,000 a 24,999
- ○ 12 25,000 a 29,999
- ○ 13 30,000 a 34,999
- ○ 14 35,000 a 39,999
- ○ 15 Over 40,000
- ○ 99 Not answered

G2. How much of that income was from AGRICULTURAL EMPLOYMENT (U.S. earnings only)? [READ/SHOW CHOICES. MARK ONE RESPONSE.]

- ○ 1 Under 500
- ○ 2 500 a 999
- ○ 3 1,000 a 2,499
- ○ 4 2,500 a 4,999
- ○ 5 5,000 a 7,499
- ○ 6 7,500 a 9,999
- ○ 7 10,000 a 12,499
- ○ 8 12,500 a 14,999
- ○ 9 15,000 a 17,499
- ○ 10 17,500 a 19,999
- ○ 11 20,000 a 24,999
- ○ 12 25,000 a 29,999
- ○ 13 30,000 a 34,999
- ○ 14 35,000 a 39,999
- ○ 15 Over 40,000
- ○ 99 Not answered

G3. What was your FAMILY'S TOTAL INCOME last year in U.S. dollars (U.S. earnings only)? [READ/SHOW CHOICES. MARK ONE RESPONSE.]

- ○ 1 Under 500
- ○ 2 500 a 999
- ○ 3 1,000 a 2,499
- ○ 4 2,500 a 4,999
- ○ 5 5,000 a 7,499
- ○ 6 7,500 a 9,999
- ○ 7 10,000 a 12,499
- ○ 8 12,500 a 14,999
- ○ 9 15,000 a 17,499
- ○ 10 17,500 a 19,999
- ○ 11 20,000 a 24,999
- ○ 12 25,000 a 29,999
- ○ 13 30,000 a 34,999
- ○ 14 35,000 a 39,999
- ○ 15 Over 40,000
- ○ 99 Not answered

Appendix A

[FOR OFFICE USE ONLY] | 3 | 4 | | | |
FARMWORKER ID

Ver. 2

PROTECTIVE EQUIPMENT

NT1 In the last 12 months, with your current (FW) employer, have you used any of the following protective equipment? [SHOW LAMINATED SHEET. CHECK ALL THAT APPLY.]

- ☐ None
- ☐ b. Gloves type 1 (cloth)
- ☐ c. Gloves type 2 (thin/light rubber)
- ☐ d. Gloves type 3 (thick/heavy rubber)
- ☐ e. Sleeves
- ☐ f. Suit
- ☐ g. Boots
- ☐ h. Respirator
- ☐ i. Hard hat
- ☐ j. Goggles
- ☐ k. Paper mask
- ☐ l. Bandana/Handkercheif
- ☐ m. Other: _____

TRAINING OR INSTRUCTIONS

NT2 Has anyone given you training or instructions in the safe use of pesticides through: video, audio cassette, classroom lecture, written material, informal talks or by any other means?

a. ...in the last 12 MONTHS, while working for your current employer?
- ☐ 0 No
- ☐ 1 Yes [SKIP TO NT3]

b. ...in the last 12 MONTHS, other than with your current employer?
- ☐ 0 No
- ☐ 1 Yes [SKIP TO NT3]

c. ...in the last 5 YEARS (but not the last 12 months)?
- ☐ 0 No [SKIP TO NT8]
- ☐ 1 Yes
- ☐ 7 Don't know [SKIP TO NT8]
- ☐ 9 Not answered [SKIP TO NT8]

NT3 ¿How was the training or instructions delivered? [READ OPTIONS AND CHECK ALL THAT APPLY.]
- ○ a. By video
- ○ b. By audio-cassette
- ○ c. Through a (formal) class/lecture
- ○ d. Through written information/materials
- ○ e. Informal instructions out in the field
- ○ f. Other: _____

NT4 How long did the training or instructions last? [READ ALL CHOICES.]
- ○ 0 Less than one - half hour
- ○ 1 Half hour - one hour
- ○ 2 >1 to 3 hours
- ○ 3 >3 hours
- ○ 7 Don't know
- ○ 9 Not answered

NT5 Who trained or instructed you? [CHECK ALL THAT APPLY.]
- ○ a. Grower/foreman/crew leader
- ○ b. Contractor or staff
- ○ c. "Government agency"
- ○ d. "Insurance agency"
- ○ e. "Union"
- ○ f. Community organization
- ○ g. Other: _____

Appendix A

NT6 In what language(s) was the training/instructions delivered? [CHECK ALL THAT APPLY.]

- ○ a. English
- ○ b. Spanish
- ○ c. French
- ○ d. Creole
- ○ e. Laotian
- ○ f. Hmong
- ○ g. Vietnamese
- ○ h. Cambodian
- ○ i. Tagalog/Ilocano
- ○ j. Mixtec
- ○ k. Kanjobal
- ○ l. Other: _____

NT7 **[READ QUESTIONS. MARK ONE RESPONSE PER QUESTION.]**

Did the training or instructions cover...

a. ...how soon could you enter a field treated with pesticides?

- ☐ 0 No
- ☐ 1 Yes
- ☐ 7 Don't know
- ☐ 9 Not answered

b. ...illnesses or injuries due to pesticides?

- ☐ 0 No
- ☐ 1 Yes
- ☐ 7 Don't know
- ☐ 9 Not answered

c. ...where to go or who to contact for emergency medical care?

- ☐ 0 No
- ☐ 1 Yes
- ☐ 7 Don't know
- ☐ 9 Not answered

NT8 Have you ever received a certification card for training or instructions in the safe use of pesticides?

- ☐ 0 No [SKIP TO NT10]
- ☐ 1 Yes
- ☐ 7 Don't know [SKIP TO NT10]
- ☐ 9 Not answered [SKIP TO NT10]

NT9 When did you receive this card?

(Month) / 1 9 (Year)

NT10 In the last 12 months, with your current (FW) employer, how do you find out the appropriate time to return to the field after it has been sprayed with pesticides? [CHECK ALL THAT APPLY.]

- ☐ a. Signs are removed
- ☐ b. Another worker informs me
- ☐ c. Employer/supervisor informs me
- ☐ d. Other (specify): _____

NT11 In the last 12 months, with your current (FW) employer, has a supervisor ever told you to enter into a field sprayed by pesticides before it was time?

- ☐ 0 No
- ☐ 1 Yes
- ☐ 7 Don't know
- ☐ 9 Not answered

20

HANDLING PESTICIDES IN THE U.S.A.

NP1 Working in the U.S. (in FW), have you loaded, mixed, or applied pesticides...

a. ...in the last 12 months, working with your current employer?

- ☐ 0 No
- ☐ 1 Yes [SKIP TO NP2]

b. ...in the last 12 months (but not with your current employer)?

- ☐ 0 No
- ☐ 1 Yes [SKIP TO NP2]

c. ...in the last 5 years (but not in the last year with any employer)?

- ☐ 0 No [SKIP TO NP6]
- ☐ 1 Yes
- ☐ 7 Don't know [SKIP TO NP6]
- ☐ 9 Not answered [PASE NP6]

NP2 The last time you did this (NP1), did you use any of the following protective equipment? [SHOW LAMINATED SHEET. CHECK ALL THAT APPLY.]

- ☐   None
- ☐ b. Gloves type 1 (cloth)
- ☐ c. Gloves type 2 (thin/light rubber)
- ☐ d. Gloves type 3 (thick/heavy rubber)
- ☐ e. Sleeves
- ☐ f. Suit
- ☐ g. Boots
- ☐ h. Respirator
- ☐ i. Hard hat
- ☐ j. Goggles
- ☐ k. Paper mask
- ☐ l. Bandana/Handkercheif
- ☐ m. Other: _____

NP3 Did you become sick or have any reaction because of this work (in NP1)?

- ☐ 0 No [SKIP TO NP6]
- ☐ 1 Yes
- ☐ 7 Don't know [SKIP TO NP6]
- ☐ 9 Not answered [SKIP TO NP6]

NP4 What problems did you have? (How did it make you sick?) [CHECK ALL THAT APPLY.]

- ☐ a. Skin problems
- ☐ b. Eye problems
- ☐ c. Nausea/vomiting
- ☐ d. Headache
- ☐ e. Numbness/Tingling
- ☐ f. Other: _____

NP5 Were you sick enough to miss 4 hours (or more) of work?

- ☐ 0 No
- ☐ 1 Yes
- ☐ 7 Don't know
- ☐ 9 Not answered

NP6 Besides what I asked you already, in the last 12 months, have you ever come in contact with pesticides by (having/being)....

a. ...sprayed or blown by the wind on you?

- ☐ 0 No   ☐ 7 Don't know
- ☐ 1 Yes   ☐ 9 Not answered

b. ...spilled on you?

- ☐ 0 No   ☐ 7 Don't know
- ☐ 1 Yes   ☐ 9 Not answered

c. ...cleaning or repairing containers or equipment used for applying or storing pesticides?

- ☐ 0 No   ☐ 7 Don't know
- ☐ 1 Yes   ☐ 9 Not answered

Appendix A

[ASK NP7 TO NP9 ONLY IF THERE IS AT LEAST ONE "YES" IN NP6]

NP7 Did you become sick or have any reaction because of this incident?

- ☐ 0 No [SKIP TO NP9]
- ☐ 1 Yes
  - It occurred in... ☐ a. NP6a?
  - ☐ b. NP6b?
  - ☐ c. NP6c?
- ☐ 7 Don't know [SKIP TO NP9]
- ☐ 9 Not answered [SKIP TO NP9]

NP8 What sickness or reaction did you have? (How did it make you sick?) [CHECK ALL THAT APPLY.]

- ☐ a. Skin problems
- ☐ b. Eye problems
- ☐ c. Nausea/vomiting
- ☐ d. Headache
- ☐ e. Numbness/Tingling
- ☐ f. Other: _____

NP9 Because of this reaction, were you sick enough to miss 4 hours (or more) of work?

- ☐ 0 No
- ☐ 1 Yes
- ☐ 7 Don't know
- ☐ 9 Not answered

NP10 Since [MONTH] of [YEAR] until NOW, [MONTH] of [YEAR] (In the last 12 months), have you received any medical attention by a doctor or nurse due to pesticide exposure?

- ☐ 0 No
- ☐ 1 Yes
  - a. When?: __ __ / 1 9 __ __ (Month/Year)
  - b. Crop?: _____
  - c. Task?: _____
  - d. What physical problem(s)?: [CHECK ALL THAT APPLY.]
    - ☐ a. Skin problems
    - ☐ b. Eye problems
    - ☐ c. Nausea/vomiting
    - ☐ d. Headache
    - ☐ e. Numbness/Tingling
    - ☐ f. Other: _____
- ☐ 7 Don't know
- ☐ 9 Not answered

A20

Appendix A

SANITATION SECTION

The following questions refer to sanitation at your job with your CURRENT (FW) EMPLOYER.

Does your current employer provide...(EVERY DAY)

NS1 ... clean drinking water and disposable drinking cups?

- ☐ 0 No water, no cups [SKIP TO NS4]
- ☐ 1 Yes, water only
- ☐ 2 Yes, water and disposable cups
- ☐ 7 Don't know [SKIP TO NS4]
- ☐ 9 Not answered [SKIP TO NS4]

NS2 Do you drink it?

- ☐ 0 No
- ☐ 1 Yes [SKIP TO NS4]
- ☐ 7 Don't know [SKIP TO NS4]
- ☐ 9 Not answered [SKIP TO NS4]

NS3 Why don't you drink it? [IF ANSWER IS "I BRING MY OWN," ASK WHY? AND ENTER RESPONSE IN "OTHER".]
[CHECK ALL THAT APPLY.]

- ☐ a. Too far away
- ☐ b. Dirty
- ☐ c. Other: _____
- ☐ d. Taste bad

NS4 ...a toilet (EVERY DAY)?

- ☐ 0 No [SKIP TO NS9]
- ☐ 1 Yes
- ☐ 7 Don't know [SKIP TO NS9]
- ☐ 9 Not answered [SKIP TO NS9]

NS5 Do you use it?

- ☐ 0 No
- ☐ 1 Yes [SKIP TO NS8]
- ☐ 7 Don't know [SKIP TO NS8]
- ☐ 9 Not answered [SKIP TO NS8]

NS6 Why don't you use it?

- ☐ a. Too far away
- ☐ b. Too dirty
- ☐ c. Other: _____

NS8 ...(provide) toilet paper EVERY DAY?

- ☐ 0 No
- ☐ 1 Yes
- ☐ 2 Yes, but insufficient supply for the day
- ☐ 7 Don't know
- ☐ 9 Not answered

NS16 With your current employer, Have you ever had to "go to" use "the bathroom" in the field/"open air"?

- ☐ 0 No [SKIP TO NS9]
- ☐ 1 Yes
- ☐ 9 Not answered [SKIP TO NS9]

NS17 Why did you have "to do it" in the field/"open air"?
[CHECK ALL THAT APPLY.]

- ☐ a. "Bathroom" is too far away
- ☐ b. Other: _____

Appendix A

NS9 ...(provide) water to wash hands EVERY DAY?
- ☐ 0 No [SKIP TO NL1]
- ☐ 1 Yes
- ☐ 7 Don't know [SKIP TO NL1]
- ☐ 9 Not answered [SKIP TO NL1]

NS10 Do you use it?
- ☐ 0 No
- ☐ 1 Yes [SKIP TO NS12]
- ☐ 7 Don't know [SKIP TO NS13]
- ☐ 9 Not answered [SKIP TO NS13]

NS11 Why don't you use it? [CHECK ANSWER(S) AND SKIP TO NS13.]
- ☐ a. Too far away
- ☐ b. Other: _____

NS12 (If "Yes" in NS10) When do you use it? [CHECK ALL THAT APPLY.]
- ☐ a. Before using toilet
- ☐ b. After using toilet
- ☐ c. Before eating
- ☐ d. Before beginning work
- ☐ e. Before leaving work
- ☐ f. Other: _____

NS13 [ASK ONLY IF THERE IS A TOILET AND A PLACE TO WASH HANDS, ASK:] Is the place to wash your hands close or far from the toilet?
- ☐ 1 Close
- ☐ 2 Far
- ☐ 3 Other: _____
- ☐ 7 Don't know
- ☐ 9 Not answered

NS14 ...(provide) soap to wash your hands EVERY DAY?
- ☐ 0 No
- ☐ 1 Yes
- ☐ 7 Don't know
- ☐ 9 Not answered

NS15 ...(provide) towels to dry your hands EVERY DAY?
- ☐ 0 No
- ☐ 1 Yes
- ☐ 7 Don't know
- ☐ 9 Not answered

NL INJURY or ACCIDENTS Section |3|4|

[Interviewer:] I would like to ask you some questions about injuries or accidents that you may have had in the last 12 months. As you know, we all get minor injuries when we are doing work, but there are times when these injuries are more serious. I would like you to think about any injuries you may have had while on a farm in the US, or traveling to or from a farm in the US, in the past 12 months. I am interested in any of these injuries that caused you to do one or more of the following: unable to work for at least 4 hours, unable able to work as hard as you normally do for at least 4 hours; had to seek medical treatment; or made you take strong medicine to allow you to keep working. [INTERVIEWER:] by strong medicine, we mean something other than over the counter medications). Do you understand, or have any questions about what I am asking?

[INTERVIEWER: If the response is "no", ask for each injury or accident listed in the Injury List to make sure no injury occurred. If the response continues to be "no" injury, skip to NLc Section "Children Injury"]

NL1 Have you had any injuries that were like what I just described to you? ☐ No ↓ ☐ Yes →   How many injuries or accidents? ☐☐ →

(Codes) Injury List:

☐ a. scrape/abrasion
☐ b. bruise/contusion
☐ c. amputation/lost body part
☐ d. sprain/strain/torn ligament/traumatic rupture
☐ e. broken bone/fracture/crushed/mangled
☐ f. dislocation
☐ g. cut/laceration/puncture/stab/jab
☐ h. burn/blister/scald
☐ i. insect bite/sting
☐ j. other: _____

Now, I would like you to describe for me what happened when you were injured? [INTERVIEWER: In the next section, you must first write the injury incident number in case there is more than one. You must ask and record answers for questions NL3 through NL15.

TAKE DOWN THE INFORMATION FIRST AS A NARRATIVE AND PROBE FOR DETAILS. TO DO THE NARRATIVE: ASK PROMPT QUESTIONS AND MARK THE CHECK LIST ABOVE THE NARRATIVE SECTION TO MAKE SURE YOU HAVE NOT MISSED ANYTHING. AFTER THE NARRATIVE, ASK (IF NECESSARY) AND ENTER RESPONSES FOR QUESTIONS NL4 TO NL15 AT THE BOTTOM OF THE NARRATIVE SECTION.

You should probe for detailed responses that lead to the injury including: When did it happen? (month/year); what they were doing? (tasks involved); how did it happen? (e.g., fell from a ladder, struck by something, lifting objects, etc.); where did it happen? (e.g. field, work shed, roadway); and any other details that can help us understand what caused the injury (e.g., others involved in the injury?, etc.). Write these responses in the narrative section.

[As a reminder, after asking each prompt for the narrative, check corresponding box. Use a different form and repeat all questions for each injury incident. If you need more space use back of page.]

Use these codes for questions NL13, NL14, and NL15:

NL13
1 Community health center
2 Private medical doctor's office/private clinic
3 Healer/"curandero"
4 Hospital
5 Emergency room
6 Migrant health clinic
7 Chiropractor or naturopath's office
8 First aid at scene
9 Dentist
10 Went to home country
11 Other: _____
97 Don't know
99 Not answered

NL14
1 Paid out of my own pocket
2 Medicaid/medicare
3 Public clinic (did not charge)
4 Employer provided health plan
5 Self or family bought health plan
6 Other: _____
7 Combination of: _____
8 Billed/did not pay
9 Worker's Compensation
97 Don't know
99 Not answered

NL15
1 Can work normally now
2 Still cannot work at full capacity
3 Still receiving treatment
4 Other: _____

25 Instructions |3|4|

Appendix A

INJURY INCIDENT #: ☐

NL3 I would like to ask you questions about this injury incident. What part(s) of your body was (were) injured and what type of injury or injuries did you have in this incident? [INTERVIEWER: SHOW FIGURE. ENTER ALL THAT APPLY AND INCLUDE DETAILS IN NARRATIVE SECTION. ASK INJURIES FOR EACH BODY PART AND CHECK ALL THAT APPLY].

Body Parts: Type of injury from Injury List on previous page (p. 25):

1. ☐   a.☐ b.☐ c.☐ d.☐ e.☐ f.☐ g.☐ h.☐ i.☐ j.☐
2. ☐   a.☐ b.☐ c.☐ d.☐ e.☐ f.☐ g.☐ h.☐ i.☐ j.☐
3. ☐   a.☐ b.☐ c.☐ d.☐ e.☐ f.☐ g.☐ h.☐ i.☐ j.☐

NARRATIVE SECTION (IF YOU NEED MORE SPACE, USE BACK PAGE)
[Check list for PROMPT questions]: ☐ how? ☐ why? ☐ details? ☐ name of machinery or tools?

_____
_____
_____
_____
_____
_____
_____
_____
_____
_____
_____
_____
_____

NL4. Where?: ☐1 field ☐2 labor camp ☐3 farm building ☐4 farm roadway ☐5 public street ☐8 Other: ☐ ☐9 not answered

NL5. When?: ___/___ | NL6. At current job? ☐0 No ☐1 Yes | NL7. Doing FW? ☐0 No ☐1 Yes | NL8. Crop?: | NL9. Task?:

NL10. Protective equipment?: ☐ | NL11. Not able to work normally >4 hours?: ☐0 No ☐1 Yes | NL12. #days not able to work normally? ☐ | NL13. Where treated? (code): ☐ | NL14. How Paid? (code): ☐ | NL15. Result? (code): ☐

26 Adult

Appendix A

INJURY INCIDENT # ☐

NL3 I would like to ask you questions about this injury incident. What part(s) of your body was (were) injured and what type of injury or injuries did you have in this incident? [INTERVIEWER: SHOW FIGURE. ENTER ALL THAT APPLY AND INCLUDE DETAILS IN NARRATIVE SECTION. ASK INJURIES FOR EACH BODY PART AND CHECK ALL THAT APPLY]:

Body Parts: Type of injury from Injury List on previous page (p. 25):

1. ☐   a.☐ b.☐ c.☐ d.☐ e.☐ f.☐ g.☐ h.☐ i.☐ j.☐
2. ☐   a.☐ b.☐ c.☐ d.☐ e.☐ f.☐ g.☐ h.☐ i.☐ j.☐
3. ☐   a.☐ b.☐ c.☐ d.☐ e.☐ f.☐ g.☐ h.☐ i.☐ j.☐

NARRATIVE SECTION (IF YOU NEED MORE SPACE, USE BACK PAGE)
[Check list for PROMPT questions]: ☐ how? ☐ why? ☐ details? ☐ name of machinery or tools?

_____
_____
_____
_____
_____
_____
_____
_____
_____
_____
_____
_____
_____
_____
_____

NL4. Where?:	☐1 field ☐2 labor camp ☐3 farm building ☐4 farm roadway ☐5 public street ☐8 Other: ☐	☐9 not answered
NL5. When?: ___/___	NL6. At current job? ☐0 No ☐1 Yes	NL7. Doing FW?: ☐0 No ☐1 Yes	NL8. Crop?:	NL9. Task?:
NL10. Protective equipment?:	NL11. Not able to work normally >4 hours?: ☐0 No ☐1 Yes	NL12. #days not able to work normally? ☐	NL13. Where treated? (code): ☐	NL14. How Paid? (code): ☐	NL15. Result? (code) ☐

27 Adult

A25

Appendix A

NLc INJURY SECTION OR ACCIDENTS FOR CHILDREN UNDER 20 YEARS OLD

"I would like to ask you questions about injury incidents that may have occured to your child(ren). These questions cover the last 12 months and refer to accidents/ injuries your child may have had while on a farm or traveling to and from a farm in the U.S.A. I am interested in injuries that caused your child(ren) to do one or more of the following: unable to do normal activities for at least 4 hours; had to seek medical treatment; had to take prescribed medicine because of the injury. Do you understand, or have any questions about what I am asking?"

NLc1 Have any of your children had any injuries that were like what I just described to you? [IF "NO" SKIP TO NLc3 TO MAKE SURE] ☐ No↓ ☐ Yes → How many children had injuries or accidents? ☐

[USE ONE FORM FOR EACH CHILD WITH INJURY, IF CHILD HAS MORE THAN ONE INJURY INCIDENT, DOCUMENT THE LAST ONE]

NLc2 [Name of child from A1]: _____ Information provided by: ☐ 0 Father ☐ 1 Mother ☐ 8 Other: _____

NLc3 I would like to ask you questions about [child's name] injury incident. What parts of [child's name] body was/were injured and what type of injury or injuries did he/she have in this incident? [INTERVIEWER: READ ALL IF NEEDED, CHECK ALL THAT APPLY]:

[INTERVIEWER: SHOW FIGURE AND INCLUDE DETAILS IN NARRATIVE SECTION. FOLLOW THE SAME INSTRUCTIONS FROM SECTION "NL" ON PAGE 25]:

Body Parts: Type of injury [mark codes from Injury List on page 25, Read all if needed]

1. _____   a.☐ b.☐ c.☐ d.☐ e.☐ f.☐ g.☐ h.☐ i.☐ j.☐
2. _____   a.☐ b.☐ c.☐ d.☐ e.☐ f.☐ g.☐ h.☐ i.☐ j.☐
3. _____   a.☐ b.☐ c.☐ d.☐ e.☐ f.☐ g.☐ h.☐ i.☐ j.☐

NARRATIVE SECTION (IF YOU NEED MORE SPACE, USE BACK PAGE)
[PROMPT question check]: ☐ how? ☐ why? ☐ details? ☐ name of machinery or tools?

_____
_____
_____
_____
_____
_____
_____

NLc4. Where?: ☐ 1 field ☐ 2 labor camp ☐ 3 farm building ☐ 4 farm roadway ☐ 5 public street ☐ 8 Other: ☐ ☐ 9 Not answered

NLc5. When?: ____/____   **NLc7.** Was [child's name] doing FW?: ☐ 0 No ☐ 1 Yes   **NLc8.** Crop?: ☐   **NLc9.** Task?: ☐

NLc10. Not able to do activities normally >4 hours?: ☐ 0 No ☐ 1 Yes   **NLc13.** Where treated? (code): ☐   **NLc14.** How Paid? (code): ☐   **NLc15.** Result? (code): ☐

28 Child

Appendix A

NLc INJURY SECTION OR ACCIDENTS FOR CHILDREN UNDER 20 YEARS OLD

[USE ONE FORM FOR EACH CHILD WITH INJURY, IF CHILD HAS MORE THAN ONE INJURY INCIDENT, DOCUMENT THE LAST ONE]

NLc2 [Name of other child with injury incident]: _____ Information provided by: ☐ 0 Father ☐ 1 Mother ☐ 8 Other: _____

1. What part(s) of [child's name] body was (were) injured in this incident? [INTERVIEWER: SHOW FIGURE. ENTER ALL THAT APPLY AND INCLUDE DETAILS IN NARRATIVE SECTION FROM PAGE 25]:

NLc3 I would like to ask you questions about this injury incident. What type of injury or injuries did you have in this incident? [INTERVIEWER: READ ALL IF NEEDED, CHECK ALL THAT APPLY]:

Body Parts: Type of injury [mark codes from Injury List on page 25]:

1. _____   a. ☐ b. ☐ c. ☐ d. ☐ e. ☐ f. ☐ g. ☐ h. ☐ i. ☐ j. ☐
2. _____   a. ☐ b. ☐ c. ☐ d. ☐ e. ☐ f. ☐ g. ☐ h. ☐ i. ☐ j. ☐
3. _____   a. ☐ b. ☐ c. ☐ d. ☐ e. ☐ f. ☐ g. ☐ h. ☐ i. ☐ j. ☐

NARRATIVE SECTION (IF YOU NEED MORE SPACE, USE BACK PAGE)
[PROMPT question check]: ☐ how? ☐ why? ☐ details? ☐ name of machinery or tools?

NLc4. Where?: ☐ 1 field ☐ 2 labor camp ☐ 3 farm building ☐ 4 farm roadway ☐ 8 Other: _____ ☐ 5 public street ☐ 9 Not answered

NLc5. When?: ___/___   **NLc7.** Was [child's name] doing FW?: ☐ 0 No ☐ 1 Yes   **NLc8.** Crop?: _____   **NLc9.** Task?: _____

NLc10. Not able to do activities normally >4 hours?: ☐ 0 No ☐ 1 Yes   **NLc13.** Where treated? (code): _____   **NLc14.** How Paid? (code): _____   **NLc15.** Result? (code): _____

29 Child

Appendix A

MUSCULOSKELETAL [INTERVIEWER: FIRST, ASK ALL QUESTIONS IN FIRST COLUMN.]

During last 12 MONTHS, [From (MONTH) of (YEAR) until now, (MONTH) of (YEAR)], have you had pain or discomfort in your... (NMS1-6) →	What type of activity were you doing when this pain/discomfort began? a.	Did you have this pain/discomfort every day for a week or more? b.	How severe was this pain/discomfort? [SHOW SCALE BELOW] c.	Were you able to work normally? IF "No", ASK: How long were you unable to work normally due to this pain/discomfort? d.
NMS1. BACK? ☐ 1 Yes → ☐ 0 No	☐ FW ☐ NF ☐ NW	☐ 0 No ☐ 1 Yes →	☐ 1 A little ☐ 2 A lot ☐ 3 Unbearable	☐☐ ☐ less than 1 day ☐ days ☐ weeks ☐ months ☐ don't know
NMS2. SHOULDER? ☐ 1 Yes → ☐ 0 No	☐ FW ☐ NF ☐ NW	☐ 0 No ☐ 1 Yes →	☐ 1 A little ☐ 2 A lot ☐ 3 Unbearable	☐☐ ☐ less than 1 day ☐ days ☐ weeks ☐ months ☐ don't know
NMS3. ELBOW/ARM? ☐ 1 Yes → ☐ 0 No	☐ FW ☐ NF ☐ NW	☐ 0 No ☐ 1 Yes →	☐ 1 A little ☐ 2 A lot ☐ 3 Unbearable	☐☐ ☐ less than 1 day ☐ days ☐ weeks ☐ months ☐ don't know
NMS4. HAND/WRIST? ☐ 1 Yes → ☐ 0 No	☐ FW ☐ NF ☐ NW	☐ 0 No ☐ 1 Yes →	☐ 1 A little ☐ 2 A lot ☐ 3 Unbearable	☐☐ ☐ less than 1 day ☐ days ☐ weeks ☐ months ☐ don't know
NMS5. LEGS/FEET (Lower extremities)? ☐ 1 Yes → ☐ 0 No	☐ FW ☐ NF ☐ NW	☐ 0 No ☐ 1 Yes →	☐ 1 A little ☐ 2 A lot ☐ 3 Unbearable	☐☐ ☐ less than 1 day ☐ days ☐ weeks ☐ months ☐ don't know
NMS6. OTHER: ☐ ☐ 1 Yes → ☐ 0 No	☐ FW ☐ NF ☐ NW	☐ 0 No ☐ 1 Yes →	☐ 1 A little ☐ 2 A lot ☐ 3 Unbearable	☐☐ ☐ less than 1 day ☐ days ☐ weeks ☐ months ☐ don't know

```
|─────────────|─────────────|─────────────|
1              2              3
A LITTLE       A LOT          UNBEARABLE
```

Appendix A

RESPIRATORY

[INTERVIEWER]: The following questions refer to the last 12 MONTHS, from (MONTH) of (YEAR) until now, (MONTH) of (YEAR).

NR1 [From (MONTH) of (YEAR) until now, (MONTH) of (YEAR)],... have you had wheezing or whistling in your chest at any time?
- ☐ 0 No
- ☐ 1 Yes:

  Number of episodes in the last 12 months. ☐☐☐

- ☐ 7 Don't know
- ☐ 9 Not answered

NR2 [From (MONTH) of (YEAR) until now, (MONTH) of (YEAR)], have you had episodes when your nose was runny or stuffy?
- ☐ 0 No
- ☐ 1 Yes:

  Number of episodes in the last 12 months. ☐☐☐

- ☐ 2 Yes, always
- ☐ 7 Don't know
- ☐ 9 Not answered

NR3 [From (MONTH) of (YEAR) until now, (MONTH) of (YEAR)], have you had episodes of watery or itchy eyes?
- ☐ 0 Never
- ☐ 1 Yes:

  Number of episodes in the last 12 months. ☐☐☐

- ☐ 2 Yes, always
- ☐ 7 Don't know
- ☐ 9 Not answered

[ASK ONLY IF THERE IS A "YES" IN NR2/NR3]

NR4 Is there any season or type of crop/task when this condition [stuffy/runny nose, watery/ itchy eyes] worsens? [MARK ALL THAT APPLY.]
- ☐ 0 No, same as usual
- ☐ 1 Yes, stuffy/runny nose or watery/itchy eyes:

  [CHECK ALL THAT APPLY.]
  What Season?
  - ☐ a. Spring
  - ☐ b. Summer
  - ☐ c. Fall
  - ☐ d. Winter

  What Crop? ☐
  What Task? ☐

- ☐ 7 Don't know
- ☐ 9 Not answered

NR5 Have you coughed on most days <u>for at least three months</u>?
- ☐ 0 No
- ☐ 1 Yes
- ☐ 7 Don't know
- ☐ 9 Not answered

NR6 Have you brought (coughed) up phlegm on most days <u>for at least three months?</u>
- ☐ 0 No
- ☐ 1 Yes
- ☐ 7 Don't know
- ☐ 9 Not answered

Appendix A

SKIN (HIS)/OCCUPATIONAL HEALTH SURVEY

[INTERVIEWER: FIRST ASK ALL QUESTIONS IN FIRST COLUMN.]

The following questions regarding skin problems refer to the last 12 MONTHS, from (MONTH) of (YEAR) until now (MONTH) of (YEAR)... (DOCUMENT ONLY "Dermititis" related problems)

(In the last 12 MONTHS) Have you had any skin problems such as redness, inflammation, discoloration, or rash on your...	a. What caused it? *(e.g., Poison Ivy/oak, chemicals, etc.)*	b. The last time you had this skin problem, what were you working on?			c. The last time you had this skin problem, were you using any protective equipment? [SHOW LAMINATED CARD]						
		1. crop? (FW)	2. task? (FW)	3. activity? (NF) (NW)	1. Cloth gloves	2. Thin rubber gloves	3. Thick rubber gloves	4. Sleeve	5. Suit	6. Boots	7. Other:	8. None
NSK1 ..hands? ☐0 No → ☐1 Yes →					☐	☐	☐	☐	☐	☐	_____	☐
NSK2 ..arms? ☐0 No → ☐1 Yes →					☐	☐	☐	☐	☐	☐	_____	☐
NSK3 ...face? ☐0 No → ☐1 Yes →					☐	☐	☐	☐	☐	☐	_____	☐
NSK4 ...any other part of your body: [____] ☐0 No → ☐1 Yes →					☐	☐	☐	☐	☐	☐	_____	☐
NSK5...any other part of your body: [____] ☐0 No → ☐1 Yes →					☐	☐	☐	☐	☐	☐	_____	☐

INFECTIONS

NI1 [In the last 12 MONTHS, from (MONTH) of (YEAR) until now, (MONTH) of (YEAR)],... have you ever had diarrhea for more than three consecutive days?

- ☐ 0 No [SKIP TO NN1]
- ☐ 1 Yes
- ☐ 7 Don't know [SKIP TO NN1]
- ☐ 9 Not answered [SKIP TO NN1]

NI2 In what month did you last have diarrhea?

Month ☐☐

NI3 How many days did the diarrhea last?

☐☐ days (or fractions of days)

NI4 How many days did the diarrhea cause you to miss work for four hours or more?

☐☐ days (or fractions of days)

NI5 Did you continue to do FW while you had the diarrhea?

- ☐ 0 No
- ☐ 1 Yes
- ☐ 7 Don't know
- ☐ 9 Not answered

NI6 Did you go to a medical doctor or medical clinic because of this diarrhea?

- ☐ 0 No
- ☐ 1 Yes
- ☐ 7 Don't know
- ☐ 9 Not answered

NEUROLOGICAL

[INTERVIEWER]: The following questions refer to the last 12 MONTHS, from (MONTH) of (YEAR) until NOW, (MONTH) of (YEAR)...have you had...

NN1 headaches regularly (i.e., more than just once in a while)?

- ☐ 0 No
- ☐ 1 Yes: How many times? ☐☐
  - ☐ a. Per week
  - or
  - ☐ b. Per month
- ☐ 7 Don't know
- ☐ 9 Not answered

NN2 ...blurred vision for more than one day?

- ☐ 0 No
- ☐ 1 Yes, always
- ☐ 2 Yes, sometimes:
  - How many times? ☐☐
  - NN2a ○ 0 Per week
  - ○ 1 Per Month
  - ○ 2 Per year
  - ○ 7 Don't know
  - ○ 9 Not answered
- ☐ 7 Don't know
- ☐ 9 Not answered

NN3 ...difficulty concentrating or trouble remembering?

- ☐ 0 No
- ☐ 1 Yes
- ☐ 7 Don't know
- ☐ 9 Not answered

A31

Appendix A

INDIVIDUAL PERSONAL HEALTH HISTORY (LIFETIME)			
[INTERVIEWER: FIRST ASK ALL QUESTIONS IN FIRST COLUMN.]			
Have you ever in your whole life been told by a doctor or nurse that you have the following conditions...?	a.	b. Are you currently taking medication for this condition? [INTERVIEWER: IF "b" IS "YES", SPECIFY MEDICATION.]	c. Have you seen a doctor or nurse in the last 12 months for this condition (NH1-11)?
NH1 ...asthma?	☐ No ↓ ☐ Yes →	☐ No → ☐ Yes [____] →	☐ No ☐ Yes
NH2 ...diabetes?	☐ No ↓ ☐ Yes →	☐ No → ☐ Yes [____] →	☐ No ☐ Yes
NH3 ...high blood pressure?	☐ No ↓ ☐ Yes →	☐ No → ☐ Yes [____] →	☐ No ☐ Yes
NH4 ...tuberculosis?	☐ No ↓ ☐ Yes →	☐ No → ☐ Yes [____] →	☐ No ☐ Yes
NH5 ...heart disease?	☐ No ↓ ☐ Yes →	☐ No → ☐ Yes [____] →	☐ No ☐ Yes
NH6 ...urinary tract infections?	☐ No ↓ ☐ Yes →	☐ No → ☐ Yes [____] →	☐ No ☐ Yes
NH7 ...thyroid disease?	☐ No ↓ ☐ Yes →	☐ No → ☐ Yes [____] →	☐ No ☐ Yes
NH8 ...cancer? Specify: [____]	☐ No ↓ ☐ Yes →	☐ No → ☐ Yes [____] →	☐ No ☐ Yes
NH9 ...hepatitis?	☐ No ↓ ☐ Yes →	☐ No → ☐ Yes [____] →	☐ No ☐ Yes
NH10 ...other: [____]	☐ No ↓ ☐ Yes →	☐ No → ☐ Yes [____] →	☐ No ☐ Yes
NH11 ...other: [____]	☐ No ↓ ☐ Yes →	☐ No → ☐ Yes [____] →	☐ No ☐ Yes

CIGARETTES

NC1 Have you smoked at least 100 cigarettes in your entire life?

☐ 0 No (SKIP TO NA1)
☐ 1 Yes
☐ 2 Don't know (SKIP TO NA1)
☐ 9 Not answered (SKIP TO NA1)

NC2 About how old were you when you started smoking cigarettes fairly regularly?

☐ 0 Never smoked regularly (SKIP TO NA1)
☐ 1 Years old ☐☐
☐ 2 Other: _____

NC3 When did you last smoke cigarettes regularly?

(MONTH)   (YEAR)
☐☐ / 1 9 ☐☐

NC4 On average, about how many cigarettes did/do you smoke a day? [1 pack = 20 cigarettes]

☐ 0 Less than 1 a day
☐ 1 Cigarettes per day ☐☐

ALCOHOL CONSUMPTION

These next few questions are about the use of beer, wine, wine coolers, cocktails, or liquor, such as tequila, vodka, gin, rum, or whiskey--all kinds of alcoholic beverages people drink at meals, special occasions, or when just relaxing.

NA1 IN THE LAST MONTH, how many days per week or per month did you drink any alcoholic beverages, on average?

☐ 0 None [SKIP TO NV1]
☐ 1 Days per week ☐☐
☐ 2 Days per month ☐☐
☐ 3 Other: _____
☐ 7 Don't know/not sure
☐ 9 Not answered

NA2 A drink is 1 can or bottle of beer, 1 glass of wine, 1 can or bottle of wine cooler, 1 cocktail, or 1 shot of liquor. On the days when you drank [FROM NA1], about how many drinks did you drink on average?

☐ 1 (number of) drinks ☐☐
☐ 2 Other: _____
☐ 7 Don't know
☐ 9 Not answered

35

VIOLENCE

From (MONTH) of (YEAR) until now (MONTH) of (YEAR)... (In the last 12 months)

NV1 [DO NOT ASK THIS QUESTION, IF INTERVIEWEE IS UNDER 18 YEARS OLD] Have you been the victim of any act of violence such as being hit, slapped, pushed, shoved, punched, threatened with a weapon, assaulted, or robbed?

- ☐ 0 No [SKIP TO NQ1]
- ☐ 1 Yes
- ☐ 7 Don't know [SKIP TO NQ1]
- ☐ 9 Not answered [SKIP TO NQ1]

Please explain how it happened: _____
_____
_____
_____
_____

NV2 Where?:
- ☐ 1 at work
- ☐ 2 home
- ☐ 3 other: _____

NV3 By whom?:
- ☐ 1 Co-worker
- ☐ 2 Relative/"Family"
- ☐ 3 Unknown
- ☐ 4 Other: _____

QUALITY OF AND ACCESS TO HEALTH CARE SECTION

[INTERVIEWER]: I would like to ask you a few final questions about health care services in general (in the U.S.). You may have given me some of this information already, but I would like to make sure it is correct...

NQ1 In the last TWO YEARS have you used any type of health care services from doctors, nurses, dentists, clinics, or hospitals in the U.S.?

- ☐ 0 No [SKIP TO NQ6]
- ☐ 1 Yes
- ☐ 7 Don't know [SKIP TO NQ6]
- ☐ 9 Not answered [SKIP TO NQ6]

NQ2 The last time...was it related to your job? ("FW" or "NF")?

- ☐ 0 No
- ☐ 1 Yes, "FW"
- ☐ 2 Yes, "NF"
- ☐ 7 Don't know
- ☐ 9 Not answered

NQ3 The last time you got attention from a health care provider, where did you go (what kind of place was it)?

- ☐ 1 Community health center
- ☐ 2 Private medical doctor's office/private clinic
- ☐ 3 Healer/"curandero"
- ☐ 4 Hospital
- ☐ 5 Emergency room
- ☐ 6 Migrant health clinic
- ☐ 7 Chiropractor or naturopath's office
- ☐ 8 Dentist
- ☐ 9 Went to home country
- ☐ 10 Other: _____
- ☐ 97 Don't know
- ☐ 99 Not answered

NQ4 The last time you got attention from a health care provider, how did you find out about the provider?

- ☐ 1 Outreach worker
- ☐ 2 Friend/relative
- ☐ 3 Newspaper/radio/television
- ☐ 4 School
- ☐ 5 Community Center
- ☐ 6 Other: _____
- ☐ 7 Don't know
- ☐ 9 Not answered

NQ5 The last time you got attention from a health care provider, who paid the majority of the cost?

☐ 1 I paid the bill out of my own pocket
☐ 2 Medicaid/Medicare
☐ 3 Public clinic (did not charge)
☐ 4 Employer provided health plan
☐ 5 Self or family bought individual health plan
☐ 6 Other plan: _____
☐ 7 Combination of: _____
☐ 8 Billed/did not pay
☐ 9 Workers' Compensation
☐ 97 Don't know
☐ 99 Not answered

NQ6 When was the last time you had dental care ("saw a dentist")?

☐ 0 Never
☐ 1 Date: ☐☐ (MONTH) / 1 9 ☐☐ (YEAR)
     Where: ☐ a. USA
            ☐ b. Abroad _____
☐ 7 Don't know
☐ 99 Not answered

NQ7 [INTERVIEWER:] I would like you to think about access to medical attention in the U.S. In general, is it easy or difficult for you to get the health care you need in the U.S.?

☐ 1 Easy
☐ 2 Dificult
☐ 7 Don't know

NQ8 When you want to get health care in the U.S., what are the main difficulties you face?
[CHECK ALL THAT APPLY.]

☐ a. No transportation, too far away
☐ b. Don't know where services are available
☐ c. Health center not open when needed
☐ d. They don't provide the services I need
☐ e. They don't speak my language
☐ f. They don't treat me with respect/I don't feel welcomed
☐ g. They don't understand my problems
☐ h. I'll lose my job
☐ i. Too expensive
☐ j. Other: _____

NQ9 If you get sick or injured, where would you go to get health care?
[CHECK ALL THAT APPLY.]

☐ a. Community health center
☐ b. Private medical doctor's office/private clinic
☐ c. Healer/"curandero"
☐ d. Hospital
☐ e. Emergency room
☐ f. Migrant health clinic
☐ g. Chiropractor or naturopath's office
☐ h. Would go to home country
☐ i. Other: _____

Appendix A

ENGLISH VERSION

We are interested in knowing whether any of the following apply to you. Please be assured that no one besides us will know your response. [READ "NL1" CHOICES IF NECESSARY]

L1	**What is your current residence status?:**	**L2**	**PROGRAMS** [DO NOT READ OPTIONS]
○ 1	I am a U.S. citizen by birth [SKIP TO E41]	○ 1	Amnesty under 5 year program
○ 2	I am a naturalized U.S.A. citizen. [Ask: Before becoming a naturalized U.S.A. citizen, in what program did you apply to obtain your permanent residence? Possible answers in "L2": 1-9, 97. THEN ASK L4#1, L4#2, AND L4#3]	○ 2	Amnesty under SAW (90 day) program
		○ 3	Cuban/Haitian entrant
○ 3	Permanent resident. "Green Card" (right to reside and work in the U.S.A.) [Ask "L2": Under which program did you apply? Possible answers in "L2": 1-9, 97. THEN ASK L3, L4#1 AND L4#2]	○ 4	Spousal petition program/Family unity
		○ 5	Labor certification program
		○ 6	Registry Program
○ 4	I have a border crossing card (right to cross the border.) [Ask "L2": Under which program did you apply? Possible answers in "L2": 1-9, 97. THEN ASK L3, L4#1 AND L4#2]	○ 7	Political asylum
		○ 8	Refugee
○ 5	Pending Status (with out documents, applied, but waiting upon an oficial decision) [Ask "L2": Under which program did you apply? Possible answers in "L2": 1-9, 97. THEN ASK L3, L4#1]	○ 9	Protective status (temporary)
		○ 10	Guestworker (H2A) program
		○ 11	Student
○ 6	Undocumented (application denied/did not apply in any programs) [Possible answers: NONE. SKIP TO E41]	○ 12	Tourist
		○ 13	Border crossing card/"passport"
○ 7	Temporary resident-Non-Immigrant Visa, (Only for a temporary time) [Ask "L2": Under which program did you apply? Possible answers 10-97. THEN ASK L3, L4 #1]	○ 97	Other. Explain:
○ 8	None of the above [Ask L2, L3, L4 #1, L4#1, and L4 #2 only if it is relevent. THEN SKIP TO E41]	○ 99	Not Answered

L3 Doyouhavegeneralworkauthorization?   ☐ 0 No   ☐ 1 Yes   ☐ 7 Don't know   ☐ 9 Not answered

L4 Date that status became effective:

1. When did you apply for (the program in "L2"? : (MONTH) (YEAR) ☐☐/19☐☐	2. [Only for those that responded to "2,3, and 4" in "L1"] When did you obtain your legal status? (MONTH) (YEAR) ☐☐/19☐☐	3. [Only for those that responded to "2" in "L1"] When did you obtain your naturalization/ became a citizen? : (MONTH) (YEAR) ☐☐/19☐☐

38

A36

Appendix B

Questionnaire Location for Items in Tables

Introduction to Appendix B

The purpose of this appendix is to enable users of this publication to look up in the questionnaire the exact questions asked of the farmworkers that provided us with this data.

How to use this appendix

Appendix B contains 8 tables that correspond to sets of tables in the main document. There are 3 columns in each of these 8 tables.
They are:
   Description in table
   Variable(s) from survey instrument
   Population

"Description in table" came directly from the tables in main body of the publication. This is indicated in the name of the table. See page **B2** title "**Variable Names from Survey Instrument for Tables 4–7 Participation in Pesticide Safety Training Programs.**" This table covers tables 4 through 7.

"From survey instrument: Variable(s)" is the designation of the questions in the survey instrument and "From survey instrument: Page" is where to find it. On **page B5** in the first column it says "Did you receive training in the safe use of pesticides?" The next column says "NT2." On page A17 (the questionnaire) you will find NT2 half way down the first column and the question "Has anyone given you training or instructions in the safe use of pesticides through: video, audio cassette, classroom lecture, written material, informal talks or by any other means?"

"Population" refers to the group of farmworkers the question applies to. For example, we only asked questions about farmworker use of PPE the last time they loaded, mixed, or applied pesticides IF they loaded, mixed, or applied pesticides. See **page B5**.

Appendix B

Table 3. Variable names from survey instrument demographic and work characteristics of farmworkers

Description in table	Variable(s) from survey instrument	Population
Age	A6	Of all farmworkers
Foreign born	A7	Of all farmworkers
Years in U.S (for those foreign born)	A8	Of all farmworkers
Place of birth	A7	Of all farmworkers
Asia		
Mexico		
Other Latin America		
U.S.		
Race	B2	Of all farmworkers
White		
Black or African American		
American Indian, Alaska Native, Indigenous		
Asian /Pacific Islander		
Other		
Ethnicity (Hispanic)	B1	Of all farmworkers
Mexican		
Mexican-American		
Puerto Rican		
Other Hispanic		
Other Ethnicity		
Family status		Of all farmworkers
Nuclear family member lives in household	A4	
Marital status of farmworker	A5	
Married		
Separated/divorced/widowed		
Single		
Children	A4	
Children in household		
Non-resident children		
Total children		
Family composition	A2	
Farmworker is a parent		
Farmworker lives with parents		
Farmworker married but does not have children		
Other		
Language		Of all farmworkers
Primary Language	B5	
Spanish		
English		
Other		

continued

Table 3. Variable names from survey instrument demographic and work characteristics of farmworkers (continued)

Description in table	Variable(s) from survey instrument	Population
Language (continued)		
Ability to read English	B6	Of those who do not answer "English" in [B5]
Not at all		
A little		
Somewhat		
Well		
Ability to speak English (for those whose primary language is not English)	B7	Of those who do not answer "English" in [B5]
Not at all		
A little		
Somewhat		
Well		
Education		Of all farmworkers
Highest grade completed	A9	
Participation in Adult education	B3	
Income		Of all farmworkers
Family income below federal poverty level	G3	
Percentage of farmworkers by family income categories (U.S. earnings)	G3	
<$500		
$500–$999		
$1,000–$2,499		
$2,500–$4,999		
$5,000–$7,499		
$7,500–$9,999		
$10,000–$12,499		
$12,500–$14,999		
$15,000–$17,499		
$17,500–$19,999		
$20,000–$24,999		
$25,000–$29,999		
$30,000–$34,999		
$35,000–$39,999		
$40,000+		
Immigration status	L1	Of all farmworkers
Citizen		
Green card		
Unauthorized		
Work authorization		

continued

Appendix B

Table 3. Variable names from survey instrument demographic and work characteristics of farmworkers (continued)

Description in table	Variable(s) from survey instrument	Population
Legal application	L1, L2	Of all farmworkers
Legalization applicant		
Family program		
Other authorization		
Unauthorized		
Citizen by birth		
Work characteristics		Of all farmworkers
Years in farm work	B10	
Hourly wage	D5-D8, D11-D18	
Number of weeks spent abroad	C6	
Number of weeks doing farm work in U.S.	C6	
Number of weeks doing non-farmwork in U.S.	C6	
Number of weeks not working in U.S.	C6	
Hours worked per week in farm work	D4	
Work for grower	C15	
Work for farm labor contractor	C15	
Method of payment	D11	
Hourly		
By piece		
Salary		
Combination of hourly and by piece		
Housing	D33A	Of all farmworkers
Farmworker rents from non-employer		
Employer provides free housing for farmworker		
Farmworker owns the house		
Farmworker rents from employer		
Employer provides free housing for farmworker and his/her family		
Farmworker rents from government or other institution		
Farmworker receives free housing from government or other institution		
Method of transportation to work	D37	Of all farmworkers
Carpool		
Drive car		
Labor bus		
Public transportation		
Walk		
Other		

Appendix B

Tables 5–8 Variable names from survey instrument participation in pesticide safety training programs

Description in table	Variable(s) from survey instrument	Population
Did you receive training in the safe use of pesticides?	NT2	Of all workers
• Received some pesticide training during the last 12 months		
With your current employer, during the last 12 months	NT2	Of all workers
With former employer, during the last 12 months	NT2	Of all workers
• No pesticide training in the last 12 months but did receive training in the last 5 years	NT2	Of all workers
• No pesticide training any time during the last 5 years	NT2	Of all workers
• How was the training delivered?		
Informal (informal instructions in the field)	NT3	Of workers trained in last 5 yrs
Formal (video, audio, written material, class)	NT3	Of workers trained in last 5 yrs
• How long was the training or instructions?		
<½ hour	NT4	Of workers trained in last 5 yrs
½ hour–1 hour	NT4	Of workers trained in last 5 yrs
>1 hour	NT4	Of workers trained in last 5 yrs
• Who trained or instructed you?		
Grower or grower's staff	NT5	Of workers trained in last 5 yrs
Farm labor contractor or farm labor contractor's staff	NT5	Of workers trained in last 5 yrs
Government agency	NT5	Of workers trained in last 5 yrs
Insurance company	NT5	Of workers trained in last 5 yrs
Other	NT5	Of workers trained in last 5 yrs
• In what language(s) was the training or instructions delivered?		
English only	NT6	Of workers trained in last 5 yrs
Spanish only	NT6	Of workers trained in last 5 yrs
Other language	NT6	Of workers trained in last 5 yrs
Bilingual English/Spanish	NT6	Of workers trained in last 5 yrs
• Was training in worker's primary language?	NT6, B5	Of workers trained in last 5 yrs
• Did the training cover the following topics required by EPA's Worker Protection Standard?		
How soon you can enter a field treated with pesticides	NT7	Of workers trained in last 5 yrs
Illness or injuries due to pesticides	NT7	Of workers trained in last 5 yrs
Where to go or who to contact for emergency medical care	NT7	Of workers trained in last 5 yrs
• Did the training cover all three topics: Reentry, illness, and emergency care	NT7	Of workers trained in last 5 yrs
Did you ever receive a certification card for training or instructions in the safe use of pesticides?		
• Received a certification card for pesticide safety training	NT8	Of all workers
• Farmworkers trained in last 12 months, who received a certification card for pesticide safety training	NT2, NT8	Of workers trained in last 12 months

Appendix B

Tables 9–12. Variable names from survey instrument personal protective equipment worn by pesticide loaders, mixers, or applicators during the last pesticide-related task performed in the last 12 months

Description in table	Variable(s) from survey instrument	Population
Have you loaded, mixed, or applied pesticides in the United States in the last 12 months?	NP1	Of all farmworkers
The last time you loaded, mixed, or applied pesticides did you wear:	Hierarchy, if gloves ranked higer were used other gloves lower in the hierarchy were not considered. Hierarchy: 1.thick rubber gloves 2. thin rubber gloves 3. cloth gloves 4. no hand protection	
• Gloves 1		
None	NP2	Of all who mixed, loaded, or applied pesticides in last 5 yrs
Cloth	NP2	Of all who mixed, loaded, or applied pesticides in last 5 yrs
Thin rubber	NP2	Of all who mixed, loaded, or applied pesticides in last 5 yrs
Thick rubber	NP2	Of all who mixed, loaded, or applied pesticides in last 5 yrs
• 1 Sleeves	NP2	Of all who mixed, loaded, or applied pesticides in last 5 yrs
• 1 Suit	NP2	Of all who mixed, loaded, or applied pesticides in last 5 yrs
• 1 Respirator	NP2	Of all who mixed, loaded, or applied pesticides in last 5 yrs
• 1 Goggles	NP2	Of all who mixed, loaded, or applied pesticides in last 5 yrs

Tables 13–16. Variable names from survey instrument availability of drinking water, toilets, and hand washing facilities

Description in table	Variable(s) from survey instrument	Population
Does your current employer provide (Every day):		
Drinking water		
No water	NS1	Of all workers
No water or disposable cups	NS1	Of all workers
Toilet		
No toilet	NS4, NS8	Of all workers
No toilet or toilet paper	NS4, NS8	Of all workers
Hand washing water		
No hand washing water	NS9	Of all workers
No hand washing water, soap, or single use towels	NS9, NS14, NS15	Of all workers

Appendix B

Tables 17–20. Variable names from survey instrument estimated 12-month prevalence of health conditions and symptoms

Description in table	Variable(s) from survey instrument	Population
Musculoskeletal pain or discomfort		
In the last 12 months have you had any pain or discomfort?	NMS1, NMS2, NMS3, NMS4, NMS5, NMS6	Of all workers
• Reported pain or discomfort in the last 12 months which affected the following areas:		
Back	NMS1	Of all workers
Shoulder/neck and upper extremities	NMS2, NMS3, NMS4	Of all workers
Lower extremities	NMS5	Of all workers
Respiratory symptoms		
• Have you had wheezing or whistling in your chest at any time in the last 12 months?	NR1	Of all workers
• Have you had episodes of runny stuffy nose or watery itchy eyes?	NR2, NR3	Of all workers
• Have you coughed or brought up phlegm on most days for at least 3 months?	NR5, NR6	Of all workers
Dermatitis		
In the last 12 months have you had any skin problem such as redness, inflammation, discoloration, or rash?	NSK1, NSK2, NSK3, NSK4, NSK5	Of all workers
• Reported dermatitis in the last 12 months which affected the following areas:		
Hands and arms	NSK1, NSK2	Of all workers
Face	NSK3	Of all workers
Other including torso and legs	NSK4, NSK5	Of all workers
Gastrointestinal problem		
• Diarrhea which lasted more than 3 days	NI1	Of all workers

Tables 21–24. Variable names from survey instrument smoking and alcohol use

Description in table	Variable(s) from survey instrument	Population
Smoking status		
Current smoker	NC1, NC3	Of all workers
Former smoker (have not smoked in last 12 months)	NC1, NC3	Of all workers
Alcohol consumption		
Consumed alcohol in last month (on average)	NA1	Of all workers
Average days per month on which alcohol was consumed	NA1	Of workers who drink
Average number of alcoholic drinks consumed per month	NA2, NA1	Of workers who drink

Appendix B

Tables 25–28. Variable names from survey instrument access to and quality of health care

Description in table	Variable(s) from survey instrument	Population
Have you used health care services in the United States in last 2 years?	NQ1	Of all farmworkers
• The last time you used health care services was it:		
Related to your farm work job?	NQ2	Of those who sought health care
Related to any work	NQ2	Of those who sought health care
Did not use health care services in last 2 years	NQ1	Of all farmworkers
For those who used health care services in the last 2 years, how did you pay?		
• For those with a problem related to the farm work job?		
Paid self	NQ1, NQ2, NQ5	Of those who sought health care for a farmwork related reason
Employer provided health plan or Workers Compensation	NQ1, NQ2, NQ5	Of those who sought health care for a farmwork related reason
Other	NQ1, NQ2, NQ5	Of those who sought health care for a farmwork related reason
• For those with a problem NOT related to the farm work job?		
Paid self	NQ1, NQ2, NQ5	Of those who sought health care for a non-farmwork related reason
Employer provided health plan or Workers Compensation	NQ1, NQ2, NQ5	Of those who sought health care for a non-farmwork related reason
Other	NQ1, NQ2, NQ5	Of those who sought health care for a non-farmwork related reason
Is it easy or difficult to get the health care services you need in the United States?	NQ7	
Difficult		Of all farmworkers
Easy		Of all farmworkers
Don't know		Of all farmworkers
When was the last time you saw a dentist (In the U.S or elsewhere)? Never	NQ6	Of all farmworkers

Table 29. Variable names from survey instrument estimated prevalence of physician-diagnosed health conditions

Health condition	Variable(s) from survey instrument	Population
Have you ever been told by a doctor or nurse that you have any health condition?	NH1, NH2, NH3, NH4, NH5, NH6, NH7, NH8, NH9, NH10, NH11	
• Type of health condition		
Asthma	NH1	Of all farmworkers
Diabetes	NH2	Of all farmworkers
High blood pressure	NH3	Of all farmworkers
Tuberculosis	NH4	Of all farmworkers
Heart disease	NH5	Of all farmworkers
Urinary tract infection	NH6	Of all farmworkers
Thyroid disease	NH7	Of all farmworkers
Cancer	NH8	Of all farmworkers
Hepatitis	NH9	Of all farmworkers
Other	NH10, NH11	Of all farmworkers

Appendix C

Crops Reported by Farmworkers in Spanish and English and Crop Categories

Introduction to Appendix C

The purpose of this appendix is to list the crops that are included in each of the categories for tables 8, 12, 16, 20, 24 and 28.

How to use this appendix

Each category is highlighted and crops are listed in alphabetical order under the category heading. English is located in first column and Spanish in the second, alphabetical by first column. The table is repeated with the Spanish column first and the English column second, alphabetical by first column.

Categories include:
- Field crops
- Fruits and nuts
- Horticulture
- Vegetables
- Miscellaneous/Multiple

Miscellaneous/multiple includes all crops that are listed as miscellaneous such as tea, sod, coffee, Christmas trees, as well as when someone reports that they are working on several different crop categories while at the same job. The crop reported is the one worked while at the current job.

Appendix C

English	Español
Field crops	**Cultivos de campo**
Alfalfa	Alfalfa
Barley	Cebada
Corn	Maíz
Cotton/Cottonseed	Algodón y Semilla
Hay	Heno
Hops	Lúpulo
Linen	Lino
Millet	Mijo, Alcandia/Zahina
Multiple Field Crops	Campos de Cultivos Diversos o Múltiples
Multiple Grains	Múltiples/Diversos Clases (Tipos) de Granos
Oats	Avena
Other Oilseeds	Otras Semillas de Aceite
Peanuts	Cacahuate, Maní
Rice	Arroz
Rye	Centeno
Silage	Ensilaje, "Forraje"
Sorghum	Sorgo, Zahina
Soybeans	Soja, Soya
Sugar Beets	Azúcar de Remolacha
Sugar Cane	Caña de Azúcar
Tobacco	Tabaco
Wheat	Trigo
Fruits and nuts	**Frutas y nueces**
Almonds	Almendra
Apples	Manzana
Apricots	Albaricoque, Chabacano (Damasco)
Avocados	Aguacate, Palta
Berry/Melons, Multiple	Múltiples, Diversas Bayas/ Granos(de Café, Cereal, Etc./ Melón, Sandía)
Blueberries	Mora, Mora Azul
Bush Berries	Mora Silvestre, Bayas
Cherries	Cereza
Citrus, Multiple	Múltiples, Diversos Cítricos
Cranberries	Baya de Arándano
Currants	Pasas
Dates	Dátiles
Deciduous Fruits, Multiple	Múltiples, Diversas Frutas de Membranas
Deciduous Nuts, Multiple	Múltiples, Diversas Nueces e Membranas
Figs	Higos
Fruits, Other	Otras Frutas
Grapefruit	Toronja, Pomelo
Grapes, Raisin	Uvas para Pasas
Grapes, Table	Uvas de Mesa
Grapes, Wine	Uvas para Vino
Kiwifruit	Kiwi

continued

English	Español
Fruits and nuts (continued)	**Fruta y nueces (continuado)**
Lemons	Limón
Limes	Limón, Lima, Limoncillo
Nectarines	Nectarina
Nuts, Multiple	Nueces Múltiples, Diversas
Olives	Aceituna, Oliva
Oranges	Naranja
Peaches	Durazno, Melocotón
Pears	Pera
Pecans	Pacanas, pecanas
Pistachios	Pistacho
Plums	Ciruela
Prunes	Ciruela Pasa, "Guindón"
Strawberries	Fresa
Tangerines	Tangerina (Mandarina)
Tree Nuts, Other	Nueces, Otras Variedades
Walnuts	Nuez, Nueces de Nogal
Horticulture	**Horticultura**
Bedding Plants	Planta Ornamental
Bulbs	Bulbos y Otros Tubérculos
Cut Flowers/Florist Cut Greenery	Flores, Flores Ornamentales
Nursery Products	Producto de Vivero
Potted Floor Plants/Ornamental Plants/Flowers	Plantas/Flores en "Pote" o Maceta
Seeds	Semillas
Vegetables	**Verduras**
Arugula (Rocket, Roquette, Rugula, Rucola)	Arugula (Rocket, Roquette, Rugula, Rucola)
Asparagus	Espárrago
Artichokes	Alcachofa
Basil	Albahaca
Beans (Fresh)	Frijol, Poroto, Judía
Beets	Remolacha, Betabel, Beterraga
Broccoli	Brócoli
Brussels Sprouts	Brotes de Bruselas / Col de Bruselas
Cabbage	Repollo, Col
Cantaloupe	Melón (Cantalupo)
Carrots	Zanahoria
Cauliflower	Coliflor
Celery	Apio
Cilantro	Cilantro, Culantro
Collards/Other Leafy Greens	Col, Repollo y Otras Frondosas
Corn, Sweet	Maíz Dulce, Elote, Choclo
Cucumbers	Pepino
Eggplant	Berenjena
Food Grown Under Cover	Planta de Cultivo Cubierta
Garlic	Ajo

continued

Appendix C

English	Español
Vegetables (continued)	**Verduras (continuado)**
Grape Leaves	Hojas de Viña (Uva)
Green Onions/Shallots	Cebollita ("China"), "Escalonia", "Chalote"
Herbs	Hierba, Yerba
Kale	"Kale", Col Rizada, Tipo de Repollo
Leeks	Puerro, Poro
Lettuce	Lechuga (Todas Clases)
Melons-Honeydew	Melón (Variedad "Honedew")
Melons, Other	Melón, Otros Melones
Mustard	Mostaza
Parsley	Perejil
Onions	Cebolla
Oriental Vegetables	Verduras/Vegetales Orientales
Peas, Dry and Lentils	Lentejas
Peas, Green	Chícharos, Arvejas, Guisantes
Peppers (Sweet And Hot)	Pimiento (Pimentón)
Potatoes	Papa
Pumpkins	Zapallo, Calabaza
Radishes	Rábano
Rapin	"Rapin"
Spinach	Espinaca
Squash	Calabaza "Squash"
Sweet Potatoes and Yam	Camote
Tomatoes	Tomate
Turnips	Nabo
Vegetables, Multiple	Múltiples o Diversos Vegetales o Legumbres
Vegetables, Other	Otros Vegetales/Verduras
Watermelons	Sandía
Miscellaneous/Multiple	**Variados/Múltiples**
Aloe Vera	Savila, Salvia
Christmas Trees	Arbol de Navidad
Clove	Clavo
Coffee	Café y Cafeto
Misc. Specialty Crops	Cultivo Especiales, Misceláneos
Miscellaneous/Multiple	Mixtos, Diversos/Múltiples, Muchos
Multiple Nursery Product	Múltiples, Muchos/ Productos de Viveros
Non-Sas (Non-Seasonal Agriclutural Services)	No-Sas (Servicios agrícolas estacionales)
Not Applicable	No Aplicable
Sod	Césped, Grama, Pasto para Jardín
Tea	Té

continued

Appendix C

Español	English
Cultivos de campo	**Field crop**
Alfalfa	Alfalfa
Algodón y Semilla	Cotton/Cottonseed
Arroz	Rice
Avena	Oats
Azúcar de Remolacha	Sugar Beets
Cacahuate, Maní	Peanuts
Campos de Cultivos Diversos o Múltiples	Multiple Field Crops
Caña de Azúcar	Sugar Cane
Cebada	Barley
Centeno	Rye
Clases (Tipos) de Granos, Ensilaje, "Forraje"	Silage
Granos Múltiples/Diversos	Multiple Grains
Heno	Hay
Lino	Linen
Lúpulo	Hops
Maíz	Corn
Mijo, Alcandia/Zahina	Millet
Otras Semillas de Aceite	Other Oilseeds
Soja, Soya	Soybeans
Sorgo, Zahina	Sorghum
Tabaco	Tobacco
Trigo	Wheat
Frutas y nueces	**Fruits and nuts**
Aceituna, Oliva	Olives
Aguacate, Palta	Avocados
Albaricoque, Chabacano (Damasco)	Apricots
Almendra	Almonds
Baya de Arándano	Cranberries
Cereza	Cherries
Ciruela	Plums
Ciruela Pasa, "Guindón"	Prunes
Dátiles	Dates
Durazno, Melocotón	Peaches
Fresa	Strawberries
Higos	Figs
Kiwi	Kiwifruit
Limón	Lemons
Limón, Lima, Limoncillo	Limes
Manzana	Apples
Mora Silvestre, Bayas	Bush Berries
Mora, Mora Azul	Blueberries
Múltiples, Diversas Bayas/ Granos (de Café, Cereal, Etc./ Melón, Sandía)	Multiple Berry/Melons
Múltiples, Diversos Cítricos	Multiple Citrus
Múltiples, Diversas Frutas de Membranas	Multiple Deciduous Fruits

continued

Appendix C

Español	English
Frutas y nueces (continuado)	**Fruit and nuts (continued)**
Múltiples, Diversas Nueces e Membranas	Multiple Deciduous Nuts
Múltiples, Diversas Nueces	Multiple Nuts
Nectarina	Nectarines
Nuez, Nueces de Nogal	Walnuts
Naranja	Oranges
Otras Frutas	Other Fruits
Otras Variedades de Nueces	Other Tree Nuts
Pasas	Currants
Pera	Pears
Pacanas, pecanas	Pecans
Pistacho	Pistachios
Tangerina (Mandarina)	Tangerines
Toronja, Pomelo	Grapefruit
Uvas de Mesa	Grapes, Table
Uvas para Pasas	Grapes, Raisin
Uvas para Vino	Grapes, Wine
Horticultura	**Horticulture**
Bulbos y Otros Tubérculos	Bulbs
Flores, Flores Ornamentales	Cut Flowers/Cut Greenery
Planta Ornamental	Bedding Plants
Plantas/Flores en "Pote" o Maceta	Potted Floor Plants/Ornamental Plants/Flowers
Producto de Vivero	Nursery Products
Semillas	Seeds
Verduras	**Vegetables**
Ajo	Garlic
Albahaca	Basil
Alcachofa	Artichokes
Apio	Celery
Arugula (Rocket, Roquette, Rugula, Rucola)	Arugula (Rocket, Roquette, Rugula, Rucola)
Berenjena	Eggplant
Brócoli	Broccoli
Brotes de Bruselas / Col de Bruselas	Brussels Sprouts
Calabaza "Squash"	Squash
Camote	Sweet Potatoes and Yam
Cebolla	Onions
Cebollita ("China"), "Escalonia", "Chalote"	Green Onions/Shallots
Chícharos, Arvejas, Guisantes	Peas, Green

continued

Español	English
Verduras (Continuado)	**Vegtables (continued)**
Cilantro, Culantro	Cilantro
Col, Repollo y Otras Frondosas	Collards/Other Leafy Greens
Coliflor	Cauliflower
Espárrago	Asparagus
Espinaca	Spinach
Frijol, Poroto, Judía	Beans (Fresh)
Hierba, Yerba	Herbs
Hojas de Viña (Uva)	Grape Leaves
"Kale", Col Rizada, Tipo de Repollo	Kale
Lechuga (Todas Clases)	Lettuce
Lentejas	Peas, Dry And Lentils
Maíz Dulce, Elote, Choclo	Corn, Sweet
Melón (Cantalupo)	Cantaloupe
Melón (Variedad "Honedew")	Melons-Honeydew
Melón, Otros Melones	Other, Melons
Mostaza	Mustard
Múltiples o Diversos Vegetales o Legumbres	Multiple Vegetables
Nabo	Turnips
Otros Vegetales/Verduras	Other Vegetables
Papa	Potatoes
Pepino	Cucumbers
Perejil	Parsley
Pimiento (Pimentón)	Peppers (Sweet And Hot)
Planta de Cultivo Cubierta	Food Grown Under Cover
Puerro, Poro	Leeks
Rábano	Radishes
"Rapin"	Rapin
Remolacha, Betabel, Beterraga	Beets
Repollo, Col	Cabbage
Sandía	Watermelons
Tomate	Tomatoes
Verduras/Vegetales Orientales	Oriental Vegetables
Zanahoria	Carrots
Zapallo, Calabaza	Pumpkins

Variados/Múltiples	Miscellaneous/Multiple
Arbol de Navidad	Christmas Trees
Café y Cafeto	Coffee
Césped, Grama, Pasto para Jardín	Sod
Clavo	Clove
Cultivos Especiales/Misceláneo	Miscellaneous Specialty Crops
Mixtos, Diversos/ Múltiples, Muchos	Miscellaneous/Multiple
Múltiples, Muchos/ Productos de Viveros	Multiple Nursery Product
No Aplicable	Not Applicable
No-Sas (Servicios agrícolas estacionales)	Non-Sas (Non-Seasonal Agriclutural Services)
Savila, Salvia	Aloe Vera
Té	Tea

Appendix D

Organizations Represented in Questionnare Planning Meeting for NAWS Occuapational Health Supplement

A meeting was held by the Department of Labor and NIOSH in April of 1998 to discuss what should be included in the Occupational Health Supplement. The meeting was attended by researchers from government agencies who are experts in farmworker health. Organizations represented included the following:

- NIOSH, Division of Surveillance, Hazard Evaluations, and Field Studies
- NIOSH, Division of Safety Research
- Department of Labor, Office of the Assistant Secretary for Policy
- Department of Labor, Occupational Safety and Health Administration
- Environmental Protection Agency, Health Effects Division
- Environmental Protection Agency, Worker Protection Branch
- National Cancer Institute
- Food and Drug Administration
- Health Resources and Services Administration
- Aguirre International
- Association of Farmworker Opportunity Programs
- Center for Public Health and Research Evaluation

Appendix E

Detailed Sample Selection Process

Goals of the Sampling Process

The first priority of the NAWS is to produce national estimates of farmworker characteristics. In 1998, the NAWS became an integral part of the JTPA (Job Training Partnership ACT) 402 funding formula, and as a result, two new requirements were added. Essentially, the JTPA provides job placement and skill development free of charge to eligible persons (JTPA). In order to successfully provide information to the JTPA, the NAWS needed to provide accurate regional estimates of JTPA eligibility, in addition to turnover and time in residence. In order to accomplish this, the sample size was increased from 2,500 to 4,000 and the original questionnaire was expanded. In the fall of 1998, after these two modifications were applied, data was collected on farmworker injuries, health, and safety.

In light of the changes to the NAWS, there were now two critical goals for the sampling process for cycles 32-34. The first was to select a random sample of farmworkers that would be nationally representative. The second was to ensure sampling of both small and large farms to adequately reflect injuries. One crucial element in attaining these goals was that the NAWS be capable of combining 2 to 4 years of data to produce regional estimates for its 12 sampling regions.

Hierarchy of Sampling Sites

Since 1998 to the present day, there have been four relevant levels of sampling sites for the NAWS, which include the region, the cluster, the county, and the employer. The regions are 12 geographic locations whose boundaries are based on USDA regions, and the clusters are groupings of counties with similar farm labor usage patterns.

Regions

Because of the importance of regional estimates for the JTPA 402 program, all regions have been included for all cycles. In the past, during the winter, regions that were fairly active were sampled individually, while relatively inactive regions were combined into a single region called "Rest of Country" from which samples were drawn. At the time this practice was started, the winter cycle or the "down cycle" began in January and lasted for 6 weeks. Because of the difficulty in finding workers for all regions, regional estimates were not of concern during this time. However, at the start of Cycle 32, the winter cycle was changed to a spring cycle, which starts in February and terminates at the end of April. Given that the "down cycle" now extended through the end of April, it became possible to find workers in all regions, no matter how far north their location. Consequently, beginning with Cycle 33, we returned to sampling individual regions in the "down cycle," rather than sampling from one large, combined group as had been done previously. The USDA Quarterly Agricultural Labor Survey (QALS) estimates of hired and contract labor for April show ample labor in nearly all regions. Yet, if problems in finding workers in northern regions do occur for a given cycle, we have the option of extending data collection into May and then reassessing the decision for the following year.

The first level of interview allocations is based on the USDA Farm Labor Survey (FLS). The USDA collects this information 4 times per year. The survey asks farm employers to provide information about their hired labor, and in most cases, their contract labor. In California and Florida, the USDA obtains contractor

Appendix E

information from the growers and then contacts the contractors for employment specifics. The USDA's purpose in the QALS is to obtain quarterly estimates of wages and employment levels regionally. This is the only data series that obtains farmworker information that is both temporal and spatial.

The USDA provides the NAWS with quarterly estimates of hired and contract employment for each of the 12 NAWS regions. Quarterly estimates are pro-rated to correspond to the three NAWS cycles. In essence, these numbers form the backbone of the cyclical and regional interview allocations.

Based on the relative proportion of farmworkers estimated for each cycle, the national total of interviews is broken down into allocations for each of the three cycles. For federal fiscal year 1998–1999, the distribution is shown in Table E1.

Table E1. Cycle allocations derived from USDA FLS

	Total	Fall	Spring	Summer
Pct of FLS	100%	34%	24%	42%
Sites	120	41	28	50
Interviews	4000	1370	949	1681

Next, on the basis of the FLS data, each region's share of workers (percentage) for all three cycles is calculated. This number is then multiplied by the total number of interviews for that cycle to produce an interview allocation.

Table E2. Regional distribution of workers and interviews for FY 1998–1999

Region	Estimated percentage of farmworkers*			Interview allocation			Totals	
	Fall	Spring	Summer	Fall	Spring	Summer	Interviews	% Interviews
Appalachia	13%	6%	9%	172	61	159	392	10%
California	29%	37%	34%	393	354	572	1319	33%
Cornbelt, Northern Plains	15%	11%	12%	210	104	196	510	13%
Delta Southeast	7%	8%	8%	103	72	128	303	8%
Florida	6%	11%	4%	76	105	72	253	6%
Lake	5%	4%	4%	75	33	75	183	5%
Mountain I,II	4%	3%	6%	57	24	106	187	5%
Mountain III	2%	3%	3%	34	29	50	113	3%
Northeast I	3%	3%	4%	40	30	61	131	3%
Northeast II	3%	3%	3%	41	27	56	124	3%
Pacific	8%	7%	9%	110	66	147	323	8%
Southern Plains	4%	5%	4%	59	45	59	163	4%
Total	100%	100%	100%	1370	950	1681	4001	100%

* Based on prior year information.

Clusters

Eighty clusters form a roster from which sampling locations are selected. These clusters are aggregates of counties that have similar farm labor usage and are roughly similar in size.

As mentioned previously, it was decided to explicitly include all regions for each cycle in order to assure adequate regional coverage. As a result, clusters were selected within regions using probabilities proportional to size (PPS).

The size measure is an estimate of the amount of farm labor in the cluster during the cycle. This measure is based on the hired and contract labor expenses from the 1992 CoA, which was the most recent CoA available at the time the sample was drawn. The CoA labor expenses are seasonally adjusted using seasonality estimates that identify the percentage of labor expenses falling into the fall, spring, and summer NAWS cycles.

To ensure an adequate number of clusters in each region, an iterative procedure was used. For each round of cluster selection, one cluster is drawn randomly from each region using the PPS measure. In successive rounds, additional counties are similarly drawn so long as the proportion of labor in previously selected clusters does not exceed a specified criterion number. The criterion number is a proportion that is sufficient to ensure that the number of clusters selected meets the number of clusters allocated to that cycle.

Table E3. Clusters per region for Fall 1998, Spring 1999 and Summer 1999 sample

Region	Total clusters	Fall 1998	Spring 1999	Summer 1999
Appalachia	6	2	2	3
California	14	7	5	7
Cornbelt, Northern Plains	8	4	3	6
Delta Southeast	7	4	3	5
Florida	11	5	3	7
Lake	5	3	2	3
Mountain I,II	4	2	1	2
Mountain III	4	2	1	2
Northeast I	4	3	2	3
Northeast II	4	3	2	3
Pacific	7	4	2	5
Southern Plains	6	3	2	4
Total	80	42	28	50

Interviews were allocated to the selected clusters proportional to the amount of seasonal farm expenses in the clusters. For example if there were two clusters and one was twice the size as the other, then the larger cluster received two-thirds of the interviews and the smaller cluster only one-third.

Counties

Within each selected cluster, one county was drawn using PPS of the county's farm labor expenses. The size measures for county selection were not seasonally adjusted. As discussed in "Calculation of seasonality estimates for fall 1998," data are insufficient at the moment to calculate seasonality estimates at the county level. It was therefore assumed that all counties in the cluster would have similar seasonality measures.

It is the NAWS policy to select one county and then collect as much of the interview allocation as possible in that county. In almost all cases, interview quotas are filled in the first county selected. Sometimes, because of unusual patterns of labor usage or disruption of labor patterns due to acts of God and government, the NAWS is unable to fill the quota in the county initially selected. In such cases, it is the policy to select an additional county from the remaining counties in the cluster using PPS. In the unlikely event that an insufficient number of farmworkers are found in the second county, supplementary counties would be selected in a similar manner.

Appendix E

Table E4. Clusters, counties and interview allocations for Fall 1998

Region	Cluster1	Interviews	County
AP	NC-5a	80	Northampton, NC
AP	TN-1	92	Lake, TN
CA	CA-5a	19	Sonoma, CA
CA	CA-8a	29	Tehama, CA
CA	CA-F	77	Fresno, CA
CA	CA-K	81	Kern, CA
CA	CA-LA	37	Orange, CA
CA	CA-M	92	Monterey, CA
CA	CA-T	58	Tulare, CA
CBNP	IL-2b	69	Kankakee, IL
CBNP	IN-2a	40	Howard, IN
CBNP	MO-2a	41	Howard, MO
CBNP	OH-2a	60	Lucas, OH
DLSE	AR-2a2	25	Cross, AR
DLSE	GA-3a	26	Grady, GA
DLSE	LA-1a	27	Iberville, LA
DLSE	MS-1c	24	Warren, MS
FL	FL-2b2	8	Polk, FL
FL	FL-3a1	11	Collier, FL
FL	FL-3b	11	Hendry, FL
FL	FL-3e	12	Dade, FL
FL	FL-PB	35	Palm Beach, FL
LK	MI-1c	28	Ottawa, MI
LK	MI-4	31	Lapeer, MI
LK	MN-1a	16	Redwood, MN
MN12	ID-3	43	Canyon, ID
MN12	MT-1	14	Richland, MT
MN3	AZ-5	11	Maricopa, AZ
MN3	AZ-6b	23	Yuma, AZ
NE1	MA-1	9	Hampden, MA
NE1	NY-2	21	Orange, NY
NE1	NY-5b	10	Chautauqua, NY
NE2	NJ-1b	11	Monmouth, NJ
NE2	PA-1a	13	Cumberland, PA
NE2	PA-1b2	18	Berks, PA
PC	OR-6M	17	Marion, OR
PC	WA-1a	19	Benton, WA
PC	WA-1c	18	Franklin, WA
PC	WA-3	57	Yakima, WA
SP	TX-10a	26	Hidalgo, TX
SP	TX-2b	19	Hale, TX
SP	TX-6b	13	Fort Bend, TX

Table E5. Clusters, counties and interview allocations for Spring 1999

Region	County Cluster	Interview	County Name
DLSE	AR-2a2	26	Crittenden, AR
DLSE	AR-2b2	25	Lonoke, AR
MN3	AZ-6b	29	Yuma, AZ
CA	CA-8a	32	Glenn, CA
CA	CA-F	88	Fresno, CA
CA	CA-K	93	Kern, CA
CA	CA-LA	47	Los Angeles, CA
CA	CA-M	93	Monterey, CA
FL	FL-2a1	17	Orange, FL
FL	FL-3b	18	Hendry, FL
FL	FL-PB	70	Palm Beach, FL
MN12	ID-3	24	Canyon, ID
CBNP	IL-1a	17	Hancock, IL
CBNP	IL-2b	50	Grundy, IL
DLSE	LA-1a	20	Iberia, LA
LK	MI-1c	20	Ottawa, MI
LK	MI-4	13	Gratiot, MI
AP	NC-5a	28	Halifax, NC
NE1	NY-2	18	Dutchess, NY
NE1	NY-5b	12	Chautauqua, NY
CBNP	OH-2a	37	Huron, OH
PC	OR-6M	20	Marion, OR
NE2	PA-1a	9	Adams, PA
NE2	PA-1b2	18	Berks, PA
AP	TN-1	33	Fayette, TN
SP	TX-10a	31	Hidalgo, TX
SP	TX-6b	14	Fort Bend, TX
PC	WA-3	46	Yakima, WA

Table E6. Clusters, counties and interview allocations for Summer 1999

Region	County Cluster	Interviews	County Name
DLSE	AR-2a2	26	Craighead, AR
DLSE	AR-2b2	25	Lee, AR
MN3	AZ-5	24	Maricopa, AZ
MN3	AZ-6b	26	Yuma, AZ
CA	CA-4a	23	San Mateo, CA
CA	CA-5a	27	Mendocino, CA
CA	CA-F	135	Fresno, CA
CA	CA-K	130	Kern, CA
CA	CA-LA	43	Los Angeles, CA
CA	CA-M	140	Monterey, CA
CA	CA-T	74	Tulare, CA
MN12	CO-0c	14	El Paso, CO
FL	FL-2b1	5	Hillsborough, FL
FL	FL-2b2	7	Osceola, FL
FL	FL-3a1	7	Collier, FL
FL	FL-3b	8	Hendry, FL
FL	FL-3c2	5	Martin, FL
FL	FL-3e	12	Dade, FL
FL	FL-PB	28	Palm Beach, FL
DLSE	GA-3a	28	Colquitt, GA
MN12	ID-3	92	Canyon, ID
CBNP	IL-1a	23	
CBNP	IL-2b	47	Bureau, IL
CBNP	IN-2a	31	Blackford, IN
NE1	MA-1	13	Franklin, MA
LK	MI-1c	28	Ottawa, MI
LK	MI-4	28	Bay, MI
LK	MN-1a	19	Cottonwood, MN
CBNP	MO-2a	22	Carroll, MO
DLSE	MS-1a	21	Coahoma, MS
DLSE	MS-1c	28	Holmes, MS
AP	NC-5a	77	Edgecombe, NC
AP	NC-7	21	Hoke, NC
NE2	NJ-1b	16	Mercer, NJ
NE1	NY-2	31	Columbia, NY
NE1	NY-5b	17	Cattaraugus, NY
CBNP	OH-2a	62	Fulton, OH
PC	OR-6M	28	Marion, OR
PC	OR-6b	21	Clackamas, OR
NE2	PA-1a	17	Adams, PA
NE2	PA-1b2	23	Berks, PA
CBNP	SD-2	11	Clay, SD
AP	TN-1	61	Crockett, TN
SP	TX-10a	14	Hidalgo, TX
SP	TX-2b	25	Cochran, TX
SP	TX-4	10	Collin, TX
SP	TX-6b	10	Brazoria, TX

continued

Appendix E

Table E6. Clusters, counties and interview allocations for Summer 1999 (continued)

Region	County Cluster	Interviews	County Name
DLSE	AR-2a2	26	Craighead, AR
DLSE	AR-2b2	25	Lee, AR
MN3	AZ-5	24	Maricopa, AZ
MN3	AZ-6b	26	Yuma, AZ
CA	CA-4a	23	San Mateo, CA
CA	CA-5a	27	Mendocino, CA
CA	CA-F	135	Fresno, CA
CA	CA-K	130	Kern, CA
CA	CA-LA	43	Los Angeles, CA
CA	CA-M	140	Monterey, CA
CA	CA-T	74	Tulare, CA
MN12	CO-0c	14	El Paso, CO
FL	FL-2b1	5	Hillsborough, FL
FL	FL-2b2	7	Osceola, FL
FL	FL-3a1	7	Collier, FL
FL	FL-3b	8	Hendry, FL
FL	FL-3c2	5	Martin, FL
FL	FL-3e	12	Dade, FL
FL	FL-PB	28	Palm Beach, FL
DLSE	GA-3a	28	Colquitt, GA
MN12	ID-3	92	Canyon, ID
CBNP	IL-1a	23	
CBNP	IL-2b	47	Bureau, IL
CBNP	IN-2a	31	Blackford, IN
NE1	MA-1	13	Franklin, MA
LK	MI-1c	28	Ottawa, MI
LK	MI-4	28	Bay, MI
LK	MN-1a	19	Cottonwood, MN
CBNP	MO-2a	22	Carroll, MO
DLSE	MS-1a	21	Coahoma, MS
DLSE	MS-1c	28	Holmes, MS
AP	NC-5a	77	Edgecombe, NC
AP	NC-7	21	Hoke, NC
NE2	NJ-1b	16	Mercer, NJ
NE1	NY-2	31	Columbia, NY
NE1	NY-5b	17	Cattaraugus, NY
CBNP	OH-2a	62	Fulton, OH
PC	OR-6M	28	Marion, OR
PC	OR-6b	21	Clackamas, OR
NE2	PA-1a	17	Adams, PA
NE2	PA-1b2	23	Berks, PA
CBNP	SD-2	11	Clay, SD
AP	TN-1	61	Crockett, TN
SP	TX-10a	14	Hidalgo, TX
SP	TX-2b	25	Cochran, TX
SP	TX-4	10	Collin, TX

continued

Table E6. Clusters, counties and interview allocations for Summer 1999 (continued)

Region	County Cluster	Interviews	County Name
SP	TX-6b	10	Brazoria, TX
PC	WA-1a	22	Benton, WA
PC	WA-3	59	Yakima, WA
PC	WA-6	17	Skagit, WA

Employers

To achieve a simple random sample of growers without specifying the number of growers to be sampled, we randomly sorted the grower lists for each selected county. Field staff contacted the growers in the order on the list and attempted to secure interviews before moving down the list.

Employers are not selected using PPS because health and safety incidents being sampled may be more frequent at small farms than large farms. Thus, PPS may skew the sample towards large farms with higher numbers of workers, so employers are selected using simple random sampling.

It is not possible to know beforehand exactly how many growers must be contacted to fill the sampling quota. Grower refusals and a variety of reasons for disqualification affect the number of growers needed to contact to fill a specific allocation of farmworkers.

Farmworkers

Farmworkers are selected at farms using the following algorithm:

Table E7. Interview Allocation

Interviews Allocated	Maximum number of Interviews per Grower
Less than 25	5
26–40	8
41-75	10
76 or more	12

Calculation of Seasonality Estimates for Fall 1998

Since we implemented a new roster of clusters in fall 1998, we had to develop new seasonality estimates for this fiscal year. In the past, seasonality estimates have been constructed as a weighted approximation of two estimates. The first estimate is obtained from farm labor experts, who are primarily agricultural extension agents. The second is constructed from Bureau of Labor Statistics (BLS) employer information on employment size. Estimates of the amount of farmwork covered by (UI) have formerly been used to construct the weight to average the BLS information and the farm expert information. When there is near universal coverage, the BLS data is the preferred estimate; nevertheless, UI coverage of farmworkers does vary considerably across states, which is why the weighted average is used. It has had a dramatic impact on how heavily farm expert information is weighted, which is directly related to the amount of UI coverage. For instance, in places where UI coverage is near universal, the farm expert information has virtually no weight. In contrast, for places where UI coverage is low, the farm expert data gets a large weight.

We were able to obtain new BLS information for our expanded counties, and were also able to commission a survey of extension agents. However, several obstacles prevented the possibility of a weighted average of BLS and farm expert information. Because of an oversight in the cooperative agreement with BLS, the BLS data obtained during summer 1998 did not include quarterly payroll information. Without payroll information, it is difficult to calculate what proportion of labor is covered by UI. Hence, we were not successful in obtaining the information needed to construct

Appendix E

a weighted average of BLS and farm expert information. The farm expert information was also more difficult to obtain than expected, and unfortunately was not available in time for sampling. Nonetheless, it will be available during fall 1998.

Because of these obstacles, the fall 1998 seasonality estimates were constructed from BLS data alone. Fortunately, in all but one state, there were adequate numbers of employers participating in the UI system to construct seasonality estimates. These seasonality estimates were made by aggregating the reported monthly employment for each month included in the corresponding NAWS cycle (e.g., June, July, August, and September for the summer cycle.). The percentage of employment corresponding to each cycle became that cluster's seasonality estimate. Once constructed, these estimates were reviewed to see whether they conformed to regional patterns and had face validity. In all cases, they did. In addition, the number of employers contributing to the estimates was reviewed. Only three of the clusters had less than 30 employers. The lowest number of employers reporting in these three clusters was 18.

State of Massachusetts declined the BLS request to provide UI data to the NAWS. So, Massachusetts estimates were derived by averaging the seasonality estimates for the remaining clusters in the region, which consisted of those in Maine and New York. The list of clusters, the number of BLS farms, and the seasonality estimates are contained in the next table.

Table E8. Seasonality estimates for NAWS county clusters for FY 1998-1999

Region	State	Cluster[1]	Number of Farms	Percent of labor		
				Fall	Spring	Summer
AP	KY	KY-2	23	34%	32%	34%
AP	NC	NC-1	18	30%	34%	36%
AP	NC	NC-5a	158	30%	27%	42%
AP	NC	NC-7	72	31%	31%	38%
AP	TN	TN-1	95	34%	30%	36%
AP	VA	VA-2	44	35%	30%	35%
CA	CA	CA-2	637	26%	37%	37%
CA	CA	CA-3b	676	27%	40%	33%
CA	CA	CA-3c	385	30%	33%	37%
CA	CA	CA-4a	522	34%	29%	37%
CA	CA	CA-5a	1171	31%	30%	38%
CA	CA	CA-7a	330	28%	29%	43%
CA	CA	CA-8a	1383	30%	27%	43%
CA	CA	CA-F	3378	28%	27%	45%
CA	CA	CA-K	1022	29%	27%	44%
CA	CA	CA-LA	1093	31%	35%	34%
CA	CA	CA-M	603	31%	27%	42%
CA	CA	CA-MD	628	28%	29%	43%
CA	CA	CA-ST	1420	30%	26%	45%
CA	CA	CA-T	1705	33%	29%	38%
CBNP	IL	IL-1a	65	29%	26%	45%

[1] The clusters follow the following naming conventions. The first term is the State. The second term is one of the following, the first letters or abbreviation of the county name in the case of a cluster composed of a single county or a letter number combination corresponding to the Employment and Training Administration (ETA) farm labor maps. See "Geographic sampling units used in the Farm Labor Area roster drawn October 1997."

continued

Table E8. Seasonality estimates for NAWS county clusters for FY 1998-1999 (continued)

Region	State	Cluster[1]	Number of Farms	Fall	Spring	Summer
CBNP	IL	IL-2b	187	32%	29%	39%
CBNP	IN	IN-2a	258	32%	28%	41%
CBNP	KS	KS-4a	78	33%	32%	35%
CBNP	MO	MO-2a	145	35%	31%	34%
CBNP	NE	NE-0f	173	34%	32%	34%
CBNP	OH	OH-2a	198	28%	21%	51%
CBNP	SD	SD-2	39	32%	33%	35%
DLSE	AL	AL-0a	70	33%	34%	34%
DLSE	AR	AR-2a2	126	35%	30%	36%
DLSE	AR	AR-2b2	167	32%	31%	37%
DLSE	GA	GA-3a	114	34%	29%	37%
DLSE	LA	LA-1a	149	42%	27%	31%
DLSE	MS	MS-1a	139	34%	28%	38%
DLSE	MS	MS-1c	250	34%	28%	38%
FL	FL	FL-0c	102	32%	36%	32%
FL	FL	FL-2a1	239	36%	36%	28%
FL	FL	FL-2b1	323	35%	47%	18%
FL	FL	FL-2b2	515	38%	42%	19%
FL	FL	FL-2c2	294	35%	40%	25%
FL	FL	FL-3a1	113	43%	38%	19%
FL	FL	FL-3a2	93	33%	40%	27%
FL	FL	FL-3b	119	42%	43%	15%
FL	FL	FL-3c2	52	26%	46%	28%
FL	FL	FL-3e	392	35%	39%	25%
FL	FL	FL-PB	344	38%	42%	21%
LK	MI	MI-1c	128	26%	31%	43%
LK	MI	MI-4	221	32%	25%	43%
LK	MN	MN-1a	170	28%	28%	44%
LK	WI	WI-0b	90	35%	27%	38%
LK	WI	WI-3b	97	37%	28%	35%
MN12	CO	CO-0c	59	32%	32%	36%
MN12	CO	CO-7a	67	29%	26%	45%
MN12	ID	ID-3	274	24%	21%	55%
MN12	MT	MT-1	68	31%	31%	37%
MN3	AZ	AZ-3	28	34%	23%	43%
MN3	AZ	AZ-5	435	30%	33%	37%
MN3	AZ	AZ-6b	222	46%	38%	16%
MN3	NM	NM-3d	49	44%	8%	48%
NE1	MA	MA-1	0	35%	27%	38%

[1] The clusters follow the following naming conventions. The first term is the State. The second term is one of the following, the first letters or abbreviation of the county name in the case of a cluster composed of a single county or a letter number combination corresponding to the Employment and Training Administration (ETA) farm labor maps. See "Geographic sampling units used in the Farm Labor Area roster drawn October 1997."

continued

Appendix E

Table E8. Seasonality estimates for NAWS county clusters for FY 1998-1999 (continued)

Region	State	Cluster[1]	Number of Farms	Fall	Percent of labor Spring	Summer
NE1	ME	ME-2a	65	35%	27%	38%
NE1	NY	NY-2	251	36%	27%	38%
NE1	NY	NY-5b	123	31%	29%	40%
NE2	DE	DE-1	105	33%	29%	38%
NE2	NJ	NJ-1b	216	33%	30%	37%
NE2	PA	PA-1a	186	37%	29%	35%
NE2	PA	PA-1b2	162	33%	34%	32%
PC	OR	OR-6b	191	25%	26%	49%
PC	OR	OR-6M	301	21%	22%	56%
PC	WA	WA-1a	347	28%	25%	47%
PC	WA	WA-1c	816	27%	27%	47%
PC	WA	WA-2a	89	27%	29%	45%
PC	WA	WA-3	1929	31%	24%	45%
PC	WA	WA-6	777	27%	27%	46%
SP	TX	TX-0d	186	35%	33%	32%
SP	TX	TX-10a	510	35%	38%	28%
SP	TX	TX-11	146	51%	22%	27%
SP	TX	TX-2b	1167	29%	22%	49%
SP	TX	TX-4	551	33%	33%	34%
SP	TX	TX-6b	336	32%	34%	34%

[1] The clusters follow the following naming conventions. The first term is the State. The second term is one of the following, the first letters or abbreviation of the county name in the case of a cluster composed of a single county or a letter number combination corresponding to the Employment and Training Administration (ETA) farm labor maps. See "Geographic sampling units used in the Farm Labor Area roster drawn October 1997."

Grower Lists

States such as California where there is near universal UI coverage, the BLS list contains almost all the agricultural employers that will be identified in that particular state. In such States, many growers who use farm labor contractors to procure harvest labor also have direct hire employees who perform other operations on the farm or ranch.

The agricultural workforce is concentrated on farms with more than 10 workers. In most States, employers are required to pay UI if they have 10 or more workers (on at least one day in each of 20 different weeks in the current or immediately preceding calendar year), or exceed a minimum payroll size ($20,000 in the current or preceding calendar quarter), otherwise they are not required to participate [DOL, 2002]. In these areas, considerable effort is made to identify agricultural employers through obtaining lists and by contacting grower organizations, local and state officials, cooperative extension agents as well as anyone who works with farmworkers or their employers [DOL, 2000].

Appendix F

Definitions of Terms

Variables not mentioned in this section came directly from the questionnaire see **Appendix B**, Origin of Data from Questionnaire.

Terms

Crop categories
Crops are grouped into five categories and include: field crops, fruits and nuts, vegetables, horticulture and miscellaneous/multiple. Each type of crop is placed in the appropriate category and is listed in **Appendix C**. This document follows other Department of Labor NAWS publications in using the term Horticulture. This category would be considered "Nursery and other Floriculture" according to the North American Industry Classification System (NAICS). "Miscellaneous/Multiple" is used when the farmworker is working on more than one crop in his/her current job.

Farmworker (for the purposes of this survey)
Workers performing crop agriculture [all crops included in the SIC code 01][1]. As defined by the USDA crop work includes "field work" in the vast majority of nursery products, cash grains, and field crops, as well as in all fruits and vegetables. Crop agriculture also includes the production of silage and other animal fodder. The NAWS population consists of nearly all farmworkers in crop agriculture, including field packers, and supervisors, and even those simultaneously holding non-farm jobs. However, the survey excludes secretaries and mechanics, H-2A temporary farmworkers, and unemployed agricultural workers. Farmworkers who have not worked in agriculture in the 14 days preceding being asked to participate are ineligible for the survey.

Fiscal year
This is a term used to describe financial allocations by the Federal government. Each year, budgets are passed that commence on the first of October and expire on the final day of September. This is what is known as the "Federal fiscal year." Fiscal years vary by state, but for the purpose of this study, the Federal fiscal year was used.

Follow-the-crop farmworker
A farmworker who has had more than one U.S. farm job and the jobs have been more than 75 miles apart. This assumes that they would have to establish a temporary domicile at or near the second job site. Follow-the-crop farmworkers can be either U.S.- or foreign-born.

Left family members behind
Not settled, farmworker is parent or farmworker is married and they are not accompanied.

Migrant status
Migrant status is defined by whether a farmworker moves for employment and how often s/he does this. For this we have established four categories, newcomer, follow-the-crop, shuttle migrant, and settled farmworker. Categories were determined from the work grid portion of the questionnaire. Categories are mutually exclusive and hierarchical with workers first being classified as "Newcomer," followed by "Follow-the-crop" farmworker, "Shuttle" farmworker, and finally "Settled" farmworker.

Newcomer
A farmworker who was born outside the United States and said they entered the United States in the year preceding the interview. This also implicates that they were excluded from this category if they had any farm work, non-work, or non-farm work period in the U.S. for 12 months or more preceding the interview.* *(According to the work grid. **See page A-7, Survey Instrument**).

Number of farmworkers employed on farm
This is the number of hired farmworkers employed on the farm at the time of the interview. The interviewer asks this of the farm operator before the farmworkers are contacted to participate in the survey.

Poverty level
Poverty determination in the NAWS is based on the US Census Bureau's method of determining poverty level using family size, family income, and poverty thresholds set by the US Census Bureau, adjusted annually.

Pesticides
For the purposes of this survey pesticides are "chemicals used to kill insects, rodents, plant diseases, and weeds."

Settled farmworker
A farmworker who does not move to find agricultural employment and spent less than 28 consecutive days abroad during the 12 months prior to the interview. If they spend more than 28 consecutive days abroad they are considered "shuttle" migrants. Settled farmworkers can be either U.S. or foreign-born.

Shuttle farmworker
A farmworker who moves once for agricultural employment during the year then returns to a "home base" to live for the remainder of the year and may work at some other job but not in agriculture. (If they did work in agriculture, they would be considered "follow-the-crop"). Shuttle farmworkers can be either U.S. or foreign-born.

Smoking status
Smoking status was determined using two steps. First, the respondents were asked if they had smoked 100 cigarettes in their lifetime **[question NC1, page A33]**. If the answer was yes, then the date they reported last smoked regularly **[question NC3, page A33]** from the interview date was used to determine their smoking status. Those who had smoked in the previous 12 months were considered to be current smokers. Those who had not were considered to be former smokers.

Stratification
The process of or result of separating a sample into several different sub-samples (or strata) according to specified criteria, such as years of farm work, migrant status, crop category or number of farmworkers on a farm.

Years of farm work
"Years of farmwork" comes directly from the questionnaire **[question B11, page A7]**. This is the number of years of farm work in the United States and includes any year in which 15 or more days were worked.

www.ingramcontent.com/pod-product-compliance
Lightning Source LLC
Chambersburg PA
CBHW080257180526
45167CB00006B/2562